Body Mechanics & Conditioning

Photo by Alinari

POMONA

Fundamentals of Body mechanics & Conditioning

AN ILLUSTRATED TEACHING MANUAL

By **MABEL LEE, B.S., LL.D.**

Professor of Physical Education and Director of Physical Education for Women, University of Nebraska

And

MIRIAM M. WAGNER, B.A., M.A.

Assistant Professor and Director of Health and Physical Education, University of Kansas City

GREENWOOD PRESS, PUBLISHERS
NEW YORK

Reprinted by permission
of W. B. Saunders Company

First Greenwood Reprinting 1969

Library of Congress Catalogue Card Number 75-91765

SBN 8371-2417-4

PRINTED IN UNITED STATES OF AMERICA

To

WILLIAM SKARSTROM, M.D.

and To The Memory of

CARL OSCAR LOUIS COLLIN, M.D.

Former Professors at Wellesley College

Who Transformed The Practice of Gymnastics and

The Study of Kinesiology Into Keen Educational Adventures

And Who

Gave Constant Inspiration For

Our Best Thinking And Our Best Work

PREFACE

Efficient body mechanics and developmental exercises, as well as adequate nutrition, medical check-ups, and recreational sports, are of major importance in safeguarding national fitness whether it be for war or for peace. Trained leadership is needed in all of these fields. To help meet this need the authors present this book in the fundamentals of body mechanics and conditioning. They hope that it will prove of practical value both to the student in training and to the inexperienced teacher who finds herself in need of a "refresher."

The material suggested in this book has been thoroughly tried out by the authors, who have spent many years in the study of the principles of mechanics, kinesiology, physiology of exercise, the art and principles of gymnastic teaching, the use of exercises for remedial measures, and the relation of exercise programs to total life experiences.

This book is written for the teacher in training who needs to learn how to sift and organize into a unified course the great amount of material available on body mechanics and conditioning, who needs experience in arranging a course with adequate emphasis and timing and with correct placement of divisions of subject matter, who needs guidance in the teaching methods peculiar to the activities of such a course, and who needs to learn how to give her pupils an understanding of the subject matter in its relation to everyday living.

The need for a course in the fundamentals of body mechanics and conditioning is discussed in Chapters I through III; subject matter for such a course is offered in Chapters IV through VIII; methods of offering the material, of evaluating class work and of safeguarding pupils are presented in Chapters IX and X; plans for organizing and conducting the course as a whole are taken up in Chapter XI; and suggested lesson plans are presented in Chapter XII.

It is the hope of the authors that this book will free its readers from the drag of the deadly details of building a course in the fundamentals of body mechanics and conditioning. May it offer quick solution to many teaching problems, thus freeing its readers to dream and hence to grow.

The authors particularly wish to thank Dr. Edmund Jacobson for permission to adapt to group class work his techniques on passive relaxation, Drs. Edward C. Schneider and Peter V. Karpovich for permission to quote and make adaptations from their text on the physiology of exercise, and the many students who have patiently given of their time for experiments in the subject matter of the text.

MABEL LEE

MIRIAM M. WAGNER

February, 1949

CONTENTS

CHAPTER I

FITNESS AND THE GOOD LIFE

There is little a person can do to lengthen his life span, but he can determine whether that life is to be full or meager. Within his inherent capacities each person has a possible level of physical fitness at which he will live most effectively and happily. During his childhood his parents, physician, and teachers should ascertain for him his possible level and should guide him towards the attainment of that level. During early youth he should learn how to maintain this level himself, understanding fully the guidance needed from the medical profession. In later years he should frequently review his life situation, his hopes and aspirations, and, in light of these should make necessary adjustments in his living habits to attain and maintain as high a level of fitness as is desirable and compatible with his physical endowment and his life plans.

No matter what the arena of life, physical fitness increases materially the opportunities for living fully. Many people live at a level of fitness far below their capacities, making drudgery both of work and play. Others, although living more nearly to their fitness level, do not experience a rich, full life through sheer lack of what it takes physically to reach the heights.

Those who in the growing years attained the heights of physical capacity and in later years have had the stamina to maintain those heights experience a fullness of life that is a closed book to the weak of spirit. Knapsack days on stiff mountain trails coming to a climax with the breath-taking view from the summit, or adventures on the waterways of the French voyageurs where one takes turns at swinging the canoe over head for a stint at a portage and at tightening guy ropes as the tent goes up at the water's edge for the night bring a spiritual uplift that is unknown to the physically inadequate.

Few indeed are there either of men or women, and particularly women, who beyond middle age are still able to live zestfully, enjoying such delights as mountain trails, canoe trips, bicycling, and golf. Such activities are usually given over almost completely to youth, young adults, and a mere handful of men who have succeeded in maintaining fitness through advancing years. What woman who has kept herself physically fit beyond the middle years does not find herself quite alone in her own age group in the average American community when she seeks companionship in physical recreational activities that are possible only to those who have accustomed themselves to live life fully!

It was not unusual in European countries before World War II to encounter entire families enjoying together a holiday of strenuous physical activity. The authors have seen these family groups at play in Germany, Holland, Belgium, France, and England. There the children grow up expecting the father and mother, the aunts and uncles to take long hikes or bike trips with them—not outings of a few hours but adventures requiring the whole weekend. Rich or poor, in summer or winter, they crowded steamers, trains, and trams to get to their "starting places" in the woods, in the mountains, at the seashores, or along the rivers. Would that the members of American families, both old and young, might learn the joys of sharing with each other such healthful and wholesome physical recreation! It would enrich family life beyond words.

Physical Reserve Needed for Times of Stress

To be fit merely for life's enjoyments is not enough. There are periods of stress and strain to be met and lived through successfully, periods calling for extra physical strength and extra physical endurance to help fortify the hard-pressed spirit. Whether of the body or of the spirit, burdens are more easily borne if there is a physical reserve. "Build thee more stately mansions, O my soul," is meant for both the body and the spirit.

FITNESS FOR WHAT?

Some people say that desirable elements of fitness for any one individual cannot be effectively listed until we answer the question, "Fitness for what? Fitness for war? Fitness for peace?" It is futile to wait for the answers to these questions. Who knows what life will or can bring to any growing child? Each day of growth is a responsibility to go forward towards maximum fitness no matter what the future may hold. Be it war, be it peace, be it a soft life, be it a hard one, the child grown to manhood can fit only into such a niche as demands no more in capacity than he possesses. The more effective his fitness and his ability to maintain it at his peak capacity, the more will life's doors be open to him.

If the answer sought must fit the one question "Fitness for War?," there yet remain many other questions: Fitness for front line fighting? Fitness for a paratrooper? Fitness for a flyer? Fitness for an office executive? Fitness as an entertainer of troops? Fitness for a submarine chaser? Fitness to make munitions? Fitness to build ships? Fitness to donate blood? Fitness to nurse the wounded? The questions are endless.

And if the answer is "Fitness for Peace," the subsequent questions are equally endless. Fitness to work on the farm, in a factory, in an office? Fitness to run a big business? Fitness to care for the sick or to educate youth? Fitness to make a nation's laws or to enforce them? Fitness to be a homemaker and to bear and rear children?

Frequently we hear people belittling fitness programs used in one country but stemming from another. They say that before we can plan wisely we must know the answers to the questions: Fitness for a German or for an American? Fitness for the Italians or for the British? Sociological and climatic considerations may be the reason for variations in the recreation programs of various races and nationalities but as far as the physiological considerations are concerned scientific studies prove the basic racial equality and needs of man.[1]

Over and above all these questions concerning life's work are those concerning life's enrichment. Fitness to enjoy life to its fullest, physically, emotionally, or spiritually?

There can be but one answer to all these questions: Fitness to the height of capacity that can be attained within the limitations inherent within any one individual. Be it in war or peace,

---

[1]Ernst Jokl and E. H. Cluver: Physical Fitness. J. A. M. A. 116:2383 (Ma· 24) 1941.

the individual must find his place in life with his established fitness pattern. Therefore let us get on without waiting for all the answers. As we wait, children are growing and are being neglected. Let us help each child so that he will develop to his maximum. No one can ask for more; no one should ask for less. Surely a sufficient answer is this: Fitness to live life to its fullest within one's special capacities no matter what the pathway may be.

FITNESS DEFINED

A committee composed of physiologists, physicians, physical educators, and public health workers, with consultants appointed by the Board of Trustees of the American Medical Association, defines fitness thus:

> Fitness for living be it in the home, on the farm, in the factory, or at the front implies freedom from disease or significant deviations from normal structure and function; enough strength, speed, agility, endurance, and skill to accomplish the maximum tasks that the day may bring; and mental and emotional adjustment appropriate to the age of the individual. Physical fitness (really physical aspects of fitness) is only a part of total fitness. The limitations of fitness are determined and modified by inheritance but within these limitations daily living practices may develop and otherwise influence fitness.
>
> Suitable work, adequate nutrition, exercise, rest, relaxation, the use of preventive and therapeutic medical services, and the avoidance of excesses, including alcohol and tobacco, are all important in maintaining physical fitness.[2]

Good Health is Basic

Good health is basic for all types of fitness; the specifics are added to that. Good health as well as disease has its symptoms, a sketchy listing of which includes physical vigor, energy, enthusiasm, alertness, good emotional reactions, lack of fatigue, and capacity to enjoy both work and play. The disciple of good health has many positive virtues such as the following:

1. He sleeps sufficiently to arise refreshed.
2. He exercises vigorously and daily be it in connection with his work or his play.

---

[2]Committee Report: The Role of Exercise in Physical Fitness. J. Health & ·ys. Educ. 14:299 (June) 1943.

3. He eats and drinks wisely both as to kind and amount of food and drink.
4. He seeks sunshine and fresh air.
5. He relaxes frequently whether with or without conscious effort.
6. He plays as well as works and has some hobby.
7. He has a regular medical check-up.
8. He maintains a wide variety of interests.
9. He practices friendliness toward mankind.
10. He cherishes the faiths upon which he builds his life.

On the negative side he avoids many things such as:

1. Excesses of all kinds including overeating and the use of drugs.
2. Infections.
3. Off weight whether it be over or under.
4. Tensions.
5. Exhaustion.

There are three major things which a person can control in regard to good health, namely diet, rest, and exercise.

Specifics of Fitness

Beyond this basic of good health and its components come the specifics of fitness such as:

1. Freedom from handicapping defects.
2. Good mental hygiene.
3. Good spirits—morale.
4. Organic power and endurance.
5. Muscular strength and flexibility.
6. Coordination.
7. Motor skills to handle the body effectively for both work and play.

For these specifics, physical education plays an important role, indeed the star role. More of this later.

Basic Requirements for Keeping Fit

Keeping fit is a matter of carrying out faithfully an intelligently planned scheme of hygienic living. Preceding all other plans there should come first of all a health examination, including a check-up on one's physical capacity, followed by the correction of remediable defects and poor health practices.

Then should come the establishment of correct health habits and regular physical activity, the kind and amount determined by the findings of the examinations. As rapidly after this as is consistent with health conditions and capacities, the exercise dosage should be stepped up to a vigorous pace and held at that level.

THE NEED FOR FITNESS

For one who would lead an effective and full life physical fitness is an essential. A fortunate few seem to attain and maintain it readily, but for the general run of people it is an elusive will-o'-the-wisp and, if captured, it must be planned and worked for. The fitness problems of a nation should be studied by experts; the nature of fitness in its varying aspects should be understood; and the fitness standards for varying types and ages of people under varying conditions should be established. During World War II our national government established the Division of Physical Fitness in the Federal Security Agency. Through it the problem of fitness was attacked intelligently, enthusiastically, and cooperatively by such groups as the American Medical Association, The American Dental Association, The American Association for Health, Physical Education and Recreation, and The United States Army and Navy.

Fitness and War

No doubt thousands of American men who were declared fit for service in World War II lost their lives because of their physical inability to endure the adverse situations into which the exigencies of war plunged them. The Army and Navy did a magnificent job of making hundreds of thousands fit in a brief period, but they had to do it the hard way because the schools and the homes of the nation had failed to maintain a thorough-going program of physical fitness for all youth. Every child in America should finish high school with as high a physical competence as it is possible for him to achieve whether he is to live in a world at war or at peace. Striving for mental competence alone is not enough.

War Fitness of American Men

Almost one-third of the nine million registrants examined by the armed services in early 1943 were found to be unfit for any form of military duty.[3] However, in contrasting these figures with rejection figures of World War I, we should remember that because of improved techniques of examination and diagnosis many were rejected in 1943 who would not have been rejected in 1917.[4] But by the standards of either period too many of our youth were found wanting. Add to these facts the additional information that of those accepted all as a whole were "soft, flabby, pampered, and in need of conditioning,"[3] and we have an astounding picture of widespread national failure in the fundamental standards of good living.

Actual statistics on the lack of physical fitness of American youth are available from many sources. A typical set of figures is the following from the testing in 1940 of a sample of 1000 young men entering the University of Illinois:

13 per cent could not swim at all (across tank).
64 per cent could not swim 50 yards.
3.1 per cent could not chin themselves once.
25.9 per cent could not chin themselves five times.
24.1 per cent could not jump an obstacle waist high.
41.7 per cent could not skin the cat.

Some 14.2 per cent were judged by instructors to have very poor (soft, flabby, or undeveloped) physiques and 10 per cent had a very poor posture.[5]

Studies of servicemen who were hospitalized show that the physical conditioning which persists through emergencies must have a good foundation through years of building. The college athlete who started body-building work as a small boy and kept it up through college years recovers quickly from accidents and without serious loss of condition. But the unconditioned youth given short intensive conditioning loses it quickly when hospitalized, stays in bed a longer period than does the conditioned boy, and after release must once more go through a complete conditioning program.[6]

Colonel Theodore P. Bank, Chief of Athletics and Recreation, Special Services Division, U. S. Army, reported that a

---

[3]Leonard G. Rowntree: Education, Health, and Physical Fitness. J. Health & Phys. Educ. 14:370 (September) 1943.

[4]In fact more men were rejected in World War II than were accepted in World War I.

[5]Thomas K. Cureton: The Unfitness of Young Men In Motor Fitness. J. A. M. A. 123:69-74 (Sept. 11) 1943.

[6]Information gleaned from conversation with Dr. C. H. McCloy, Civilian Consultant of the Surgeon General, U. S. Army, Reconditioning Consultant's Division.

great many of the draftees were absolutely exhausted after doing a slow 300-yard run. He said:

> Our physical education programs in high schools and colleges have been a miserable failure. Physical education through play must be discarded and a more rugged program substituted. We must assume our share of the responsibility for the unnecessary loss of American lives. Many of our boys have perished because of the accumulation of fatigue, the lack of endurance, stamina and certain agilities. You read about the men who struggled through swamps and jungles and over mountains for days and days before reaching safety or survived in rubber boats for many days before being rescued. They had the strength and stamina to survive such ordeals, but you don't read about the hundreds that did not have such strength and stamina. They did not live to make a report.[7]

War Fitness of American Women

Thousands of American women became a direct part of the nation's war effort during World War II. In adjusting to their new role in life, most of these women found themselves "in the red" on the physical fitness side of the ledger. In the absence of statistical data on servicewomen and war workers we must fall back on the findings that are available, and unfortunately these are limited almost exclusively to the college field. When unfavorable news comes from that area there is little reason to believe that better conditions exist among American women in general.

Information on the physical fitness of entering college girls should tell us much about the American girl at her best since these girls represent both the age group and the class group that is the most favored for fitness.

One of the myths about the American girl is that she is a great sports girl. This is far from the truth. Any teacher of sports and physical education in any high school or college in the country can testify to the utter indifference to participation in sports on the part of the vast majority of American girls. Under skillful motivation to get them out for a sports-tournament or for individual participation, the endeavor is usually considered a great success if even as few as 10 to 15 per cent of the girls turn out. The American man, too, loves to take his sport "sitting down" much more than popular writers

[7]Physical Fitness from the Standpoint of the Army. Proceedings of National War Fitness Conference 1943, American Association for Health, Physical Education, and Recreation, p. 39.

would lead the world to believe. Fewer American college men turn out voluntarily for sports participation than turn out voluntarily for gymnastic drill in some foreign universities. (The men who do turn out voluntarily for gymnastic drill in this country are almost negligible.)

What are the findings on the physical fitness of the select group of American girls who go to college? A physical performance fitness test was made on 296 Radcliffe College women under the guidance of research workers at the Harvard Fatigue Laboratory. All of these young women had previously been given medical examinations and were pronounced physically fit by general standards. Yet of these young women only 22 per cent had the physical strength and endurance to complete the test (the step test).[8] Similar results were found from a great variety of tests given during the war years in colleges and universities in all parts of the country. They bespeak a more vigorous training for high school girls of the nation.

And what of Women in the Armed Forces? Figures for 1943 informed us that there were 232,000 American women in uniform at that time.[9] Official information on the physical condition of these women is meager. What little is available gives evidence that the women have made no better showing than have the men. However, the figures on the two groups are not comparable since those for men stem from the examinations of millions of draftees and those for women from the examinations of only a few thousand volunteers. However, the statistics on women in the services should be a fair sampling of the condition of American women as a whole.

The Surgeon General's office reported in the fall of 1944 a one-third rejection for the Women's Army Corps, an unusually large number of these rejections being due to excessive overweight.[10] It is interesting to note that this figure matches almost exactly the rejection percentage of men draftees as a whole.

[8]Harriet L. Clarke: A Functional Physical Fitness Test for College Women. J. Health & Phys. Educ. 14:359 (September) 1943.

[9]Dorothy LaSalle: Dance and Fitness for Women. J. Health & Phys. Educ. 14:309 (June) 1943.

[10]Information given to the National Civilian Advisory Committee of the WAC at its meeting in Washington, September, 1944. One of the authors is a member of this committee.

The same complaint of general lack of strength, flexibility, and endurance that was made against the men who were accepted for service was also made against the great majority of women who were accepted. In the early training days the accident rate was high owing to the weak musculature of the women. Lt. Stella Salvison, a member of the Illinois Committee on the Economic and Legal Status of Women for the American Association of University Women who was one of the first class of 336 officers trained for the Women's Army Corps, reported that the daily forty-minute drill and forty minutes of calisthenics showed up the physical softness of women college graduates. She was convinced that college physical education programs should include more physical education and more rigorous calisthenics for all students.[11] Information gleaned by one of the authors from personal conversations with many officers of the Women's Army Corps bears out these statements. Entrance examinations revealed much overweight, poor posture and unnecessary work fatigue.

All women accepted by the Women's Army Corps were put through a rigorous program of conditioning. Visits behind the scenes at the First WAC Training Center at Fort Des Moines gave hope for the future physical fitness training of American girls. The thousands of women who went through this training found a new fullness of life in physical well-being.[12]

With this rigorous training the WAC built up sufficient strength and endurance to be able to be at work each morning by 6 o'clock, to stay on the job all day long and to have sufficient reserve left by evening for physical recreational activities. To visit the mammoth recreation hall at Fort Des Moines and see hundreds of girls enthusiastically playing basketball and volleyball after a long day of hard work in barracks, classrooms, garages, and truck- and jeep-driving practice field was indeed an inspiration. It gave hope for an America of the

---

[11]The WAAC Officers Take Over. Journal of American Association of University Women, October, 1942.

[12]The War Department WAC Field Manual Physical Training, FM 35-20, 15 July 1943 gives a good indication of what this training encompassed. The excellent work done by Major Donna Niles, assisted by Major Brenda Boynton, in setting up this program was given special commendation through a citation from the American Academy of Physical Education to the War Department in 1945.

future with a citizenry physically efficient. We may well be proud of the physical stamina of the women of our armed services!

In training women for war the Army learned some interesting things about women as compared to men. For example, it was discovered that women cannot be pushed as hard the first thing in the morning as can men but they can hold on longer at the end of the day. Also women ate much less food than the Army thought they would. They demanded more fruit than do men but much less meat, coffee, tea, sugar, vegetables, pastries, and eggs.

And what about American women in war work? Of the many million American women in industry during World War II, over five million of them were directly connected with the war effort, making planes, ships, tanks, and ammunition. Industry had much to say of their superb performance in the face of great physical inefficiency. These jobs called for skills beyond the usual reach of the average American girl. They required much lifting, carrying, pushing, pulling, stooping, squatting, crawling, balancing, climbing, and even working on scaffolds. Frequently it was necessary for these women to climb steep ladders and at the top to swing themselves onto a catwalk.

Thorough-going body-conditioning work in the physical education program in high school would have made war work much easier and much more efficient for thousands of these women, and incidentally would have made the work progress faster. However, American women, in spite of their physical shortcomings, did a magnificent job, but a higher level of fitness for both men and women would have saved many lives in the services and many serious accidents and lost man-hours in industry.

Fitness and the Pursuits of Peace

It is not realized sufficiently that the rate of industrial and agricultural production of a people depends upon the physical efficiency of the people, and that the physical efficiency of the people has a direct bearing on the health of the nation and the rate of progress of education. Many people declared fit by standards of employability are not able to retain positions because of physical inefficiency. This physical inefficiency is usually accompanied by mental dullness, "lack of inner discipline," high incidence of minor illnesses, bad posture, and "lax general bearing" and produces a bad impression in general.[1]

Dr. Ernst Jokl, Medical Consultant of the South African Defense Force, and Dr. E. H. Cluver, Director of the South African Institute for Medical Research, conducted an experiment on a group of young men who, although pronounced fit by the usual standards, were nevertheless incapable of holding jobs or of fitting into normal life. All, in spite of organic soundness, were physically inefficient, lacking in strength and endurance, poorly nourished, and living in unfavorable surroundings. These young men were placed in a situation where they had good living conditions and good nutrition and they were given a special course of training in physical efficiency. After six months of this treatment all were "transformed from substandard youths" into normal young men capable of living effectively.[1] What an easy solution to many of life's problems —good living conditions and proper nutrition for all (a civic responsibility where the home fails) and the development of physical efficiency (a direct challenge to the schools through their physical education departments).

Modern living with its labor-saving machines and its easy access to energy-saving transportation is designed largely around a sedentary existence which makes the maintenance of physical efficiency a real problem. And yet everyone, no matter what his place in life, no matter what his type of work, needs a high degree of physical fitness for a margin of emotional as well as physical safety so that he will be able to stand up under emergencies. It is not large muscles that are needed but effectively working ones. Nor is it mere brute endurance that is needed but a will to do which must be backed up by actual physical powers of staying at a task.

Some people must earn their livelihood by physical labor, some by sedentary work; yet all must be prepared to live effectively; all need reserves of strength and endurance.

Fitness of Women For Tomorrow's World

Whether it was in the armed forces or on the industrial front, American women during World War II found a new place for themselves in our national life. From now on any catalog of careers for women must take into consideration the long list of new skills learned in industry and in the armed services. Indeed, some women are even looking forward to a career in the armed services.

In the 1930's, Margaret Culkin Banning reported that of the thirty-seven million adult women in the United States, only three million were employed. Five million of the unemployed were past sixty, leaving nineteen million who were either home-makers or merely idle. World War II materially changed those figures. In the summer of 1943, there were sixteen million women in industry alone. Fourteen per cent of these women were between the ages of fourteen and eighteen; 22 per cent were over forty-five years of age; 25 per cent were married; and 12 per cent had children under sixteen years of age. The trend of work for those women at that time was a ten-hour day for seven days a week, on the night shift, and with two to four hours a day extra time spent in getting to and from the job. Some of these women were loading freight cars and driving heavy buses and trucks, and this grind went on even during the menstrual period.[13]

Of the many millions of women who were not employed during World War II a high percentage did volunteer services. Surely most of these women have retained this aroused interest in volunteer service for the common good. Surely every city, every town, every village will have some form of sewing centers, youth centers, recreation clubs, emergency canteen services, citizenship-training centers, and many other community services that will find their existence largely through women volunteer workers. Living in a democracy should mean not only the four freedoms but the four responsibilities: (1) to procure justice for the downtrodden (2) to aid the underprivi-ledged (3) to care for the sick and (4) to protect the weak. To measure up to these tasks and at the same time to fill one's

[13]Louise Stitt: Problems of Women in Industry. Proceedings of Victory-Through-Fitness Workshop, National Association for Physical Education of College Women, June, 1943, p. 21.

own niche in life, one needs stamina—nervous, mental, physical, and emotional.

Demands of Homemaking

Homemaking with its marketing and carrying of loads, cooking, housework, washing, cleaning, mending, care for children, and a thousand and one other duties, calls for physical strength and endurance to spare so that the mother doesn't drag around at a constant fatigue level. Childbearing calls for strong abdominal muscles, flexible not tense. The avoidance of the usual run of home accidents (which claim more victims than accidents in any other category) calls for better arm and shoulder girdle strength, for better coordination, for better agility.

More Strength and Endurance Needed

Living effectively in today's world requires much physical stamina. In many ways it requires even more skill and energy than was demanded of the early settlers. Today's girl must be sturdy as well as glamorous. She must have strength, endurance, and steady nerves. Strength does not mean bulky, bulging muscles; it means muscles with more of their fibers able to share in a task.

Women have greater life expectancy than men and greater resistance to disease, but do they know as well how to live a full, rich life? They do not need to drag through life constantly tired and nervous. A bit more strength and endurance would remedy much of this. As Steinhaus[14] so well says,

> No matter what the task in life, a bit more strength in the shoulders and upper back would send most women home from the office, store, or factory at the close of day with less of shoulder, neck and back ache: indeed with none at all. Furthermore, a bit more physical endurance would send these same women home with energy to spare for an evening of fun. For most women a little more of strength, speed, agility, and endurance would make all the difference in the world between dragging around too weary to enjoy either work or play and having pep and enthusiasm for both. Beyond all this a strong body and steady nerves mean ability to meet emergencies calmly and come out on top.

---

[14]Dr. Arthur H. Steinhaus, Professor of Physiology, George Williams College, former Chief of the Division of Physical Education and Health Activities, U. S. Office of Education.

The Woman of Tomorrow

American women asked for equal suffrage and got it, for equal opportunity and in many situations now have it; now they must assume equal responsibilities. The emergency tasks of World War II equalled the challenge of pioneer days, and the tasks that lie ahead will, no doubt, carry women into a life as different from the pre-war days as those days were themselves different from the gay nineties. The popular pin-up girl of tomorrow will be a healthier, sturdier girl of steady nerves.

The "global fifties" are upon us. An increasing number of women are holding down two jobs; one the management of the home and the rearing of children, the other her share in the world's work. The few women who have done this in the past have encountered difficulties undreamed of by men who as a rule have but the one main task. As more women undertake this dual role the task will become easier, for their very numbers will force the world to become better organized to aid them.

The Middle Years

During World War II, reports came quite insistently from industry to the effect that women approaching middle age proved more stable in war work both physically and emotionally than did the younger women. Apparently the middle years need not be the health and fitness hazard most people have heretofore attributed to them. Research on the comparative strength and endurance of women in the twenty's, thirty's, and forty's needs yet to be done. What are their relative conditioning needs? What is their relative ability to recover good condition after it has been lost? How far can the women in these three different decades be pushed physically with safety?

Whatever are the specific needs of adults to maintain a well conditioned level, recreational needs and the need to maintain his previous best level of physical fitness still exist for each and every one. Therefore "keeping fit" and engaging in recreational physical activities are important for all adults who hope to continue a rich, full life into and beyond the middle years.

THE RESPONSIBILITY OF THE SCHOOLS

Since the majority of all American children of school age attend school,[15] the responsibility of education for the physical training of the rising generation is one that cannot be ignored with impunity. A child gives much of its life to the school. The school in return should give the child the training and development he will need later on to meet life's demands whether it be in the realm of the mental, emotional, social, or physical. The complete training and development which education should give to children deserves more serious consideration than it usually receives in the average school. Teachers, superintendents, boards of education, and parents quite usually agree on the need for social and mental development, but in too many instances they are naïvely unaware of their lack of concern for the child's physical development, or, if aware, they are just as naïvely satisfied that a program of athletics for the athletically inclined boys and a few simple folk dances for the girls "fill the bill." The school must face its particular responsibility for the normal physiological growth of all children under its care. Beyond that, the school must face the fact that a child's exercise and play life have direct bearing on the development of social aspects of living and that a good physical education which produces a body well developed and a personality well adjusted through rich experiences in group physical activity bears a very important relationship to the child's chances of leading a life enriched by his physical poise and emotional adjustment.

In the Lower Grades

In the lower grades where children still have a natural urge for hanging, swinging, climbing, running, and dodging, formalized exercises are not necessary to condition them, provided equipment or natural objects for these activities are available. A "jungle gym," climbing ropes, trees, stationary ladders, and

[15]Gladwin Hill in the New York Times Magazine of July 29, 1945 states that in World War I only 5 per cent of American soldiers were college men, only 15 per cent had partial high school education and only 20 per cent had grade school education, but in World War II, 13 per cent of our soldiers were college men, 51 per cent had gone to high school, and 63 per cent had attended grade school.

free space to run in are all that is needed to give young children excellent conditioning, provided there is leadership to supervise and motivate so that all take part and do so with safety.

In the Upper Grades and High School

Children in the upper grades and high school have lost much of their instinctive urge for the type of activity that is of itself conditioning, yet their physiological growth and development are still in progress. At this time rapid increase in strength outruns the development of coordinations and endurance so that the child is awkward and easily fatigued. Guarding against too much strain, coordinations and endurance should be developed as rapidly as possible.[16]

Jokl and Cluver tell us that before puberty girls have endurance almost equivalent to that of boys the same age, but at thirteen years of age while the boy's endurance is increasing the girl's endurance has reached its maximum. After that age, the girl's endurance declines so that girls of eighteen and twenty do not have as much endurance as do girls of six and seven. (How frequently have the authors noticed this disparity in endurance between these two age groups in mountain climbing!) Although girls lose in powers of endurance after thirteen years of age, they increase in strength because of their increasing body weight and muscular development.[17] In the high school program of conditioning, we must keep this fact in mind. Just how long this decline in endurance persists we do not know but the above mentioned research clearly shows that no upward swing has as yet set in by eighteen years of age. This range of thirteen to eighteen years covers the entire high school period and bespeaks a conditioning program that does not expect too much of most girls in endurance activities. If they play games as strenuous as basketball they should play for short quarters, not for long halves. The exceptionally sturdy and athletically enthusiastic girls, who are always in the minority, should not be allowed to set the

[16]For a fuller discussion of physiological and emotional needs of the growing child in relation to exercise see Mabel Lee: The Conduct of Physical Education. New York, A. S. Barnes & Co., 1937, Chapter III.

[17]Jokl and Cluver: op. cit., p. 2837.

pace for the majority. Their surplus energy and exuberance can be used to good advantage in leadership assignments. Also, neither boys nor girls should be permitted to pursue a lone sport or even a lone duet of sports even though they like these sports to the exclusion of all others. Early sports training should include as wide a variety of skills as possible, thus opening up a rich recreational life for the child grown to adulthood. Similarly, their physical training should include thorough instruction and constant practice in body mechanics and conditioning, thus opening up a future that is satisfying because the body is responsive and capable for whatever tasks life may impose.

If the grade schools would assume the responsibility of educating children physically, as well as mentally and socially, they would turn them out with well-developed bodies, with remedial body mechanics difficulties corrected, with good physical coordination, flexibility, strength, and endurance, and with a repertoire of sports, aquatics, and rhythmical skills that would materially enrich their lives. But American education on the whole is blind to these responsibilities and opportunities. Rare indeed, also, are parents who themselves provide these things for their children to make up for their schooling deficiencies.

The fact that over fifty per cent of our high school girls receive no physical training in their last two years in high school[18] should give pause for serious thought. These same girls will soon be the mothers of a new generation with important fitness needs for childbearing, homemaking, and in many cases for the dual role of child raising and breadwinning. Thousands of these young girls will bear heavy burdens. Why does the school close its eyes to its physical education responsibilities towards these girls? Every high school girl should be given instruction in body mechanics and conditioning.

In College

When the day comes when our schools actually give an adequate physical training to all school children of America the college physical education program can take on what

[18]Dorothy LaSalle: Fitness Today on the Home Front. J. Health & Phys. Educ. 15:535 (December) 1944.

seems logically to be its real function, namely, that of offering the students a rich recreational life of physical activities and a fitness-maintenance program. Dean Gildersleeve, formerly of Barnard College, has listed five values which a girl should acquire from a college education, namely, (1) the value of work so that she can contribute to the welfare of society, (2) the value of affectional relationships particularly in home life, (3) the value of spiritual and ethical character, (4) the value of participation in community, state, national, and international affairs and (5) the value of recreation. The authors would add to this list a sixth value, that of physical fitness in its full implications so that the girl may gain the utmost from the other five values.

Because of the failure of the lower schools, college physical education programs must plug away at the development of good body mechanics and conditioning and the fundamentals of sports and rhythmical skills. Fortunately, college age is not yet too late for this instruction; college students are still in the physical growth and developmental period, and according to psychologists, they are still in a period quite ripe for the learning of motor skills. Even those young people who do enter college properly trained in physical skills still have the continuing need for physical recreation and the maintenance of good physical condition.

World War II profoundly altered the thinking of college girls in regard to their physical fitness. A formerly feigned dislike for physical activity gradually turned to an earnest searching for shortcuts to fitness. These girls saw their fathers, brothers, and sweethearts, one by one, drawn into the conflict. They followed them in mind and spirit through days, weeks, and months of strenuous physical training and gradually came to realize that the physical fitness of a loved one might readily mean all the difference in the world between life and death for that loved one. Little by little came the realization of their own need to be fit and to keep fit to do their share in the war effort. After the prewar days of cynicism this awakening of American college girls was a heartening thing to observe. Suddenly the awakening flowered into requests from the girls themselves for more strenuous work in physical education classes, for exercises they could do at sorority houses and dormitories at bedtime, for increased programs, and for an understanding of fitness in relation to war, peace, and everyday life.

In the fall of 1943, the Department of Physical Education for Women at the University of Nebraska set up a Planning Board which was made up of freshmen girls who met regularly with the director of the department to talk over the girls' interests, desires, and problems in reference to the course which was required of all freshmen women, and which consisted of just such a course as the authors offer in this book with main emphasis on body conditioning and the body mechanics of everyday life. The students took an intelligent interest in the course outline and offered many constructive suggestions, most of which were built around requests for more and still more strenuous work.

In the spring of 1945, the department called for volunteers for a special course in conditioning for second year students. It was explained that the work would be very strenuous, much more strenuous than that offered even in freshmen body mechanics and conditioning and that no one should volunteer who was afraid of a hard physical workout. Girls volunteered in such large numbers that the class enrollment had to be closed even before the announcement had been made in all classes. The girls finally selected for class membership were given a special medical check at the Health Service and only those approved for strenuous activity were accepted. This select group was then given an unusually strenuous course in conditioning, and at the end of the course the girls were asked to give their frank opinions of the work. Following are a few of their remarks:

1. "It is fun, fast moving and full of excitement. Even the soreness is fun. I think there are plenty of others who would like to try it."

2. "I think it is a superb class, and I have enjoyed every minute of it."

3. "I have enjoyed this whole semester's work and would not have missed it for anything."

4. "Enjoyed this very much. Wish it would have been an entire semester. I feel I could have done better with more training."

5. "All in all, it was a lot of fun, and I think it did me good."

6. "I can't think of anything I disliked. It was hard on the muscles at times, but I asked for it when I volunteered."

7. "I enjoyed the course very much and liked all the work, and especially enjoyed comparing scores to see how physically unfit I was. I think other people would enjoy doing it, but it should be a volunteer group!"

8. "I can't exactly say why I enjoyed the course, but I really did. In fact I even looked forward to it."

9. "I believe it is one of the best courses possible and I certainly was glad to be chosen to take this course in physical education."

10. "Thanks a lot—I feel in better shape now and also have a better shape."

11. "I honestly feel this course has helped to build me up and develop my muscles as they should be—mainly in my legs and thighs. Also it seemed to place the weight where it belonged!"

12. "I sincerely feel that from this Research Project Class I have gained much that is very worthwhile. Not only has it stimulated my interest in the various types of physical exercises and activities in which we engaged, but it has also developed my powers of endurance and speed—to say the least of what it has done for me in giving me increased will-power to undertake strenuous physical tests with confidence and assurance. I am convinced that other groups would benefit from and enjoy such a class."

The results of this experiment prove quite conclusively that college girls can "take it" and are deeply interested in conditioning work that is actually strenuous enough to condition them.

Colleges all over the country are today finding enthusiastic support among the students for serious "keep fit" programs of work, and all colleges with live departments of physical education are requiring all entering freshmen to take a course in the fundamentals of body mechanics and conditioning.

CHAPTER II

THE ROLE OF EXERCISE IN FITNESS

Exercise is by no means the only performer in the fitness drama but it does have an important part to play, a part commensurate with nutrition, rest, and medical care; and the role played by exercise should be both recognized and understood.[1] It can develop organic power, strength of muscle, and endurance of the heart-lung mechanism; it can give release from tensions; and in certain ways it can aid in emotional adjustment. The person who does not use his muscles, heart, and lungs to their most efficient capacity functions below par. As Schneider says:

> It is . . . common experience to find that a man in poor physical condition is easily exhausted by mental and physical exercise; he is irritable, likely to have morbid thoughts, petty ailments, and a low morale; he may have a sallow complexion and dull eyes; and he frequently complains of constipation, headache, nervousness and insomnia. On the other hand, it is equally common to observe in a man of good physical condition evidences of mental and bodily vigor such as alertness, cheerfulness, high morale, bright eyes, elastic step, healthy complexion, and capacity for arduous mental and physical work.[2]

The role of exercise is not to develop and maintain bulky muscles but to develop and maintain efficiently working muscles and to maintain effective heart-lung mechanism. To quote Schneider again:

> Frequently repeated exercise, extending over months and years, brings about many striking alterations in the structure and function of organs of the body. It has been well said that the benefits of muscular work cannot be overestimated, that exercise is necessary for healthy existence, that it is a physiologic need of a primitive kind which cannot safely be eliminated by

[1]For much of the material in this chapter, the authors wish to acknowledge as their authority Edward C. Schneider and Peter V. Karpovich: Physiology of Muscular Activity. 3d ed. Philadelphia, W. B. Saunders Co., 1948, accepting, however, full responsibility for their interpretation of the material.

[2]Edward C. Schneider: Physiology of Muscular Activity. 2nd ed. Philadelphia, W. B. Saunders Co., 1939, p. 258.

civilization . . . It is difficult to find men who have been injured by muscular exercise but easy to find many who have failed of normal development and been ruined by lack of it.[3]

To review the definition of fitness drawn up by the Joint Committee of The American Medical Association and the American Association of Health, Physical Education, and Recreation we find that fitness implies, among other things, "enough strength, speed, agility, endurance, and skill to accomplish the maximum tasks that the day may bring." These specifics are acquired and maintained only through exercise. For many of life's tasks, and pleasures as well, it is important to be able to do strenuous activity effectively and for long periods of time. Also it is important, no matter what the nature of the work, to be able to do each day's task with but little fatigue and with some reserve for emergency demands. Beyond this it is important as an element in the development of spiritual fitness to strengthen the will by acquiring the habit of persisting in the face of difficult physical tasks.

A well-planned and well-directed program of vigorous exercise will produce these elements of fitness. There are definite physiological principles that apply to the development and maintenance of fitness through exercise. The most fundamental of these have to do with the development of strength and endurance. However, before these can be understood it is important to understand the nature of the teamwork of the muscles, heart, and lungs—the three star performers on the strength-and-endurance-developing team.

THE ROLE OF MUSCLES IN EXERCISE

All movements of the body are brought about through the action of muscles. These muscles are of various sizes and shapes, some short, some long, some flat, and some spindle shaped. Each is a bundle of fibers, each one of which is a unit in itself. Some muscles have only a few fibers and others have an innumerable number of them.

The Function of Muscle

The function of muscle is to contract, and it contracts upon signal from the nervous system. Just how all of this comes

[3]Schneider: op. cit., p. 255.

about is an uncertain story of the anatomical mechanisms involved in the transmission of nerve impulses. Indeed physiologists insist that there is as yet no adequate explanation of the mechanism of muscular contraction. For our purposes perhaps it is sufficient to say that when a muscle "works" in the ordinary sense, the individual fibers shorten their length making the muscle as a whole become shorter. Not all the fibers enter into a piece of work, however, unless it is an "all-out" effort. Let us consider a muscle as a factory. It has many workers which it can call upon as orders come in, and they are called upon in such numbers as are needed for any particular job at hand. If the order is small, only a few workers are called upon and accordingly only those few receive pay in development. If the muscle turns out nothing but small jobs most of its workers remain poor from lack of employment. However, if there is a continuous flow of small orders, first one group of workers goes on the job, then another and another until all have had a turn. The factory keeps running but with no one group of workers doing very much and no one group earning very much. But when larger orders come in more workers are called upon and the extra workers begin to function for longer periods. As long as this increased business lasts these extras keep "in the pay" but when business becomes lax again they are "laid off" and they soon become poor again. However, if the larger orders pour in regularly and crowd each other the various groups begin working together and working "overtime" until, if the large orders last over a goodly period of time, the factory pays dividends and gradually increases in power. So it is with muscles and their fibers. The amount of work assigned them in quantity, quality, and duration determines whether they will be flabby or well developed.

Mechanics of Movement

To bring about movement no one muscle acts alone. One muscle or set of muscles may be contracting to produce the movement, another to guide the direction of the movement, a third to "hold" some part of the body so the acting muscles will have a firm support to pull on, and still a fourth to stop the movement before it goes too far. Some muscles are capable of producing a speedy movement, others a strong movement.

By working together in various ways several muscles or groups of muscles may combine to produce a more powerful movement, a speedier movement, or a more controlled movement than one muscle could produce alone. This "teamwork" is the result of too intricate a mechanism to be discussed here but it produces the graceful performance of the dancer, the accuracy and steadiness of the golfer, and the speed of the sprinter.

To a person who has curiosity about the mechanics of things in general, there is no more interesting investigation than the study of the mechanics of bodily movement. Armed with a knowledge of the musculature of the human body and the physiology of muscular action, and grounded in the fundamentals of the actions of levers and the elementary principles of the mechanics of motion, the study of kinesiology (the science of the body in motion) becomes a real adventure. For the person who would intelligently devise exercises for specific purposes it is a required study. It is also a "must" for one who would direct others to highly skilled performances in any field of physical activity.[4]

Muscle Soreness

Continuous or frequently repeated "all-out" contractions of a muscle or long-continued holding of a muscle in a state of contraction cause injury to the muscle fibers, resulting in soreness and stiffness; and if the contractions have been of a violent nature they may result in tearing, stretching, or even rupturing the fibers, producing the condition known as muscle strain.

At each contraction of a muscle the blood supply to the fiber is reduced. This cutting off of the blood supply, if continuous enough, gives a sensation of pain in the muscle which is believed to be due to poor oxidation from lowered oxygen supply. Also the accumulation of waste products may produce changes in the fiber, causing a sensation of discomfort. This pain and discomfort are recognized as "muscle soreness." When a muscle works, fluid from the blood collects in it and

---

[4]Two excellent books on the study of kinesiology are Wilbur Bowen and R. Tait McKenzie: Applied Anatomy and Kinesiology. 4th ed., revised. Philadelphia, Lea & Febiger, 1928 and Gladys M. Scott: Analysis of Human Motion. New York, F. S. Crofts & Co., 1942.

after it ceases to work it may be some time before this excess fluid is carried away. If the accumulation is excessive it causes the muscle to be swollen, thus producing a sensation of stiffness.

According to the time of the onset of muscle soreness there are three types as follows: (1) that which sets in as the exercising progresses (if severe enough it will cause the person to stop the activity), (2) that which occurs immediately after some strain and (3) that which does not appear until sometime after the exercise is over, generally not until the next day. The first may be explained by the accumulation of waste products or the lack of oxygen, with recovery rapid once activity ceases. The second is due to injury to muscle fibers and the recovery is slow. The third type is not as yet satisfactorily explained by physiologists. It may persist for a number of days.[5]

Ordinary muscle soreness requires only two things: application of heat (a hot bath is fine) and continued light use of the muscles affected. Massage is helpful if the rubbing movements are toward the heart, thus helping to push the flow of venous blood away from the sore muscle. (It is the venous blood that carries away the waste products.) In fact, much muscle soreness may be avoided by taking a hot shower or hot tub bath immediately after strenuous exercise, thus stepping up the removal of waste products from the muscles and speeding up delivery of fuel to the muscle to repair injured fibers. Ordinary muscle soreness need not cause alarm. It is a normal accompaniment of strenuous muscular activity. Like ordinary fatigue it merely causes discomfort and is a signal that the muscle was not in very good condition to meet the demand made upon it.

THE ROLE OF HEART AND LUNGS IN EXERCISE

Oxygen is a gaseous substance ever present in air. Air is drawn into the lungs, and from that part of the air which gets down into the farthest pockets of the lungs (the alveoli) the oxygen is transferred to the blood stream as the blood passes through the walls of the alveoli. In the blood stream the oxygen is loosely combined with the hemoglobin of the red corpuscles and having thus "thumbed a ride" travels along in the blood

[5] Arrangement from Schneider: op. cit., pp. 37-38.

stream to the heart. The force of contraction of the heart muscle sends the oxygen along in the blood stream to all parts of the body.

When the blood stream has arrived at its farthest outposts (the capillaries) the oxygen diffuses out into the muscle cell and there it unites with the different food elements (carbohydrates, fats, and proteins) which also have been carried to the muscle in the blood stream, having been picked up in the digestive tract. This union of oxygen with the food elements results in what is called "oxidation" or a "burning" which means a breaking down into simpler chemical compounds by the addition of oxygen to a substance. Consider the muscle cell as a furnace; the food elements, the fuel; the oxygen, the air without which the fuel will not burn; and certain enzymes, which also are carried in the blood stream, as the match which ignites the fuel and sets off the combustion. This combustion results in the production of energy which recharges the muscle which, like an electric storage battery, must be recharged if it is to continue to function. Now that the muscle is charged it is in a condition to act when the "electric current" is turned on (when the nerve impulse to contract reaches it).

Also the heart and lungs function together to remove from the muscle cell the waste products of its labor, i.e., the ashes accumulated from the burning that takes place in this "furnace." These waste products (carbon dioxide) being in the form of gas diffuse through the cell wall into the blood stream, from there are diffused into the lungs, and from there are breathed out of the body.

The more a muscle works (contracts) the more oxygen it must have; and the more oxygen it must have, the more the heart and lungs must work to supply the oxygen. There is no such physiological possibility as breathing deeply over a period of time with the body at rest in order to store up a supply of oxygen in advance of its needs. As Schneider and Karpovich say, "man practically lives a hand-to-mouth existence as far as his oxygen supply is concerned".[6] In other words, the body has no storage tank for oxygen. But when an increased supply is needed to recharge the muscle because it

[6] Schneider and Karpovich: op. cit., p. 35.

has increased its work output, the lungs and heart automatically step up their activity in order to deliver the materials promptly for this faster recharging. If they are not able to respond efficiently then the muscle does not get recharged efficiently. Similarly the muscle is not relieved of its waste products efficiently.

Training which comes about through repeated experiences requiring this increased effort on the part of the heart and lungs increases the efficiency of the heart and lungs. Research proves that slow, deep breathing is the most effective rate for ordinary activity. Sixteen breaths per minute is the average adult rate.[7] A stepped-up requirement for oxygen brought on by increased activity forces a stepped-up rate of breathing. While the well-trained person handles this temporary increase of breathing rate easily, the untrained is forced to meet it with discomfort which shows up in labored breathing and undue breathlessness.

Since the heart is a muscle, it gets its workout only through its contractions but, unlike the skeletal muscles, its work is involuntary and automatic. When large muscle groups work more rapidly than usual they require a faster delivery of oxygen and a faster discharge of carbon dioxide. Fortunately, our bodies are so constructed that the order for the heart and lungs to produce faster is an automatic one. When this increased workout is experienced regularly over a long period of time the whole heart and lung mechanism gradually becomes stronger and can do increased work with less effort. In the person accustomed to regular, vigorous exercise the heart empties itself more completely at each beat than does the heart of the untrained person.[8] Therefore it is an inescapable fact that the regular vigorous exercise of the large muscle groups of the body has a direct relation to heart and lung efficiency.

Pulse Rate and Fitness

Pulse rate (heartbeat rate) increases with exercise. If the exercise is severe, this increase reaches a certain level in the trained person and remains there as the exercise is continued,

[7]Schneider and Karpovich: op. cit., p. 78.
[8]Schneider: op. cit., p. 281.

but in the untrained it continues to rise.[9] Hence care should be taken that exercise is not severe until the person has had considerable training. This training should consist of gradually stepped-up exercise.

Since a nervous condition quickens the heart rate it is highly undesirable to permit untrained persons to engage in strenuous exercise of a highly excitable nature, such as competitive sports, where there is a high stake on the need to win. The emotional excitement adds to the increased demand which the strenuous exercise itself makes upon the heart. To handle such an order safely the heart should be in a well-trained state.

The person who is physically fit shows less difference between the standing and reclining pulse than does the untrained person. The best rate for young women is a change from fifty beats per minute reclining to sixty beats per minute standing.[10]

The length of time it takes for the heart to return to its normal beat after exercise depends upon the amount and kind of work and the condition of the worker. Ordinarily the heart rate returns to normal within a minute after a half minute of step-ups but after exhaustive work it may not return to normal for several hours.[11]

Second Wind

People who have never exercised long enough to experience the physiological sensation known as "second wind" have no idea what a satisfying experience it is. Those who put the overload principle into practice experience second wind as they are adjusting to a higher level of activity. (A discussion of the overload principle follows later in this chapter.) To increase endurance one must exercise beyond this second-wind level. Just before reaching this stage discomfort is experienced, the heart is beating fast, there is distressing breathlessness, and most probably some muscle pain exists. The person feels that he will be forced to quit at any moment but he persists, and after a while he realizes that the feelings of dis-

[9]Schneider: op. cit., p. 198.

[10]A. H. Turner: The Adjustment of Heart Rate and Arterial Pressures in Healthy Young Women During Prolonged Standing. Am. J. Physiol. 81:197, 1927.

[11]Schneider and Karpovich: op. cit., p. 169.

comfort have disappeared; his heart action has slowed down; he is no longer "out of breath"; he has forgotten his aches and pains; and he feels refreshed. In fact he is refreshed, for now he has "second wind." Now he can continue with the activity as though he were just beginning it.

There are several explanations of this phenomenon. One is that in the first stage of activity there is a temporary increase in the girth of the muscle because of the accumulation of liquid pouring into the tissues from the blood and also there is congestion of blood in the arteries of the muscles. These things cause distress until the blood vessels on the outermost surfaces of the muscle open up and relieve the congestion. With this relief the discomfort disappears, the muscle is able to function better, and the delivery rate of oxygen to the muscle has become adjusted to the increased needs, thus removing the urgency of greatly increased breathing and heart pumping. Other explanations involve considerations of oxygen intake, carbon dioxide output, the cooling effect of the evaporation of water on the skin, and the rise in temperature brought on by muscular work. Research work shows that second wind is experienced earlier in warm air than in cold, and when more heavily clothed. Also the greater the rate of work, the earlier it appears.[12]

THE ROLE OF FOOD IN EXERCISE

The old theory that in muscular work the protein of the muscle fiber itself was destroyed and therefore needed to be replaced by a protein diet is now looked upon as fallacious. Research shows that for the greatest possible muscular effort the body requires carbohydrates for its source of energy. Although fats are a good source of energy, the work produced from them is performed 10 per cent less economically than that produced from carbohydrates. Trainers of marathon racers have come to realize that the eating of candy before and during a race improves muscular effort and physical condition.[13] Certain kinds of candy have been manufactured for the use of athletes, mountain climbers, and other persons who are able

[12] Schneider and Karpovich: op. cit., pp. 101-104.

[13] Ibid., pp. 28-29.

to know in advance that great muscular effort will be required of them within definite time limits. These candies are supposed to contain sugar in its predigested form to save time in body chemistry. By taking certain definite amounts of these candies at a set time before the work is to begin there will be a goodly supply of predigested sugar in the body ready for quick conversion into energy when needed. Thus the person can reach his work peak sooner and can ward off fatigue longer.

Many taboos and some erroneous sanctions have been built around food in its relation to exercise. For example, it has long been considered unwise to exercise soon after a meal, but now physiologists tell us that if the exercise produces no discomfort it is not harmful. Also, recent research has exploded the old notion that to go in swimming soon after a meal brings on muscular cramps. Then there is that old and stubbornly adhered-to theory that exercise is an excellent means of reducing. Yes, exercise will reduce a person if he will stick to it long enough. For example, it has been computed that to lose 1¼ pounds of body weight by exercising, one must do such exercises as the following: walk 66½ miles at the rate of one mile in seventeen minutes, saw wood for fourteen hours, or walk to the top of Washington Monument forty-eight times.[14] This shows the absurdity of trying to reduce by exercise. The way to reduce is to regulate the food intake. This should be done according to the advice of persons trained in the science of nutrition, or, if the weight is excessive due to glandular disturbance, it should be remedied under competent medical advice.

The capacity to develop strength and endurance is of course dependent upon the presence in the muscle cell of proper kinds and amounts of food elements as well as upon the presence of oxygen. There are many excellent sources of information available on proper nutrition for fitness.[15] To be intelligent on

---

[14]Arthur H. Steinhaus, Alma Hawkins, Charles D. Giauque and Edward Thomas: How to Keep Fit and Like It. Chicago, Consolidated Book Publishers, Inc., 1943. This little pamphlet is full of valuable material and may be purchased from the publishers or from George Williams College, Chicago.

[15]Such as Edna P. Amidon, Dorothy E. Bradbury and Vivian V. Drenckhahn: Good Food and Nutrition for Young People and Their Families. New York, John Wiley & Sons, Inc., 1946 and Fern Silver: Food and Nutrition. New York, Appleton-Century Co., 1941.

any phase of fitness one should understand the fundamentals of other phases, since all are interdependent. Indeed, futile and foolhardy are any attempts to develop muscular strength and endurance in a person who does not eat nourishing food and get sufficient rest and sleep. Therefore teachers responsible for the physical-conditioning program of children should have some check on their nutritional status.

THE DEVELOPMENT OF STRENGTH AND ENDURANCE

Engineers plan their structures so that they can offer resistance to many times the stresses they will actually need to bear. So the body engineer should plan for more strength and endurance than may ever be actually needed.

Muscle Tone

Before entering upon a discussion of the development of strength and endurance in muscles, a few words should be said about muscle tone, that state in a muscle in which some fibers are constantly in contraction giving it a quality of firmness. Good muscle tone is essential to the building of strength and endurance. Muscles of good tone do not show fatigue as readily as do those of poor tone. Also, they can swing into action quickly because there is but little slack to be taken up. When a long railroad train pulls out of a station, the engine is going at an appreciable speed before the rear coach is set into motion. This is because of the slack that needs to be taken up before the inertia of each coach can be overcome and before the coaches can act as one unit following the pull of the engine. The less slack to be taken up, the sooner all parts swing into action. So it is with muscles.

A muscle that is well nourished and well exercised has good tone. This good tone can be lost not only through poor diet and lack of sufficient exercise but also through a dissipation of its energy by too much and too long-continued holding in a state of either stretching or contraction.

The Overload Principle

All exercise does not develop strength and endurance in a muscle. It is only when a muscle works with greater intensity than normally or when it works for a longer period of time than normally that it acquires more strength and endurance. Development occurs in a muscle only when it is being pushed beyond its usual performance—in other words, is "overloaded." Recognition of this physiological fact has given rise to the tenet on muscular development known as "the overload principle." It means just what its name implies, i.e., that there must be an overloading, a pushing beyond normal limits if a muscle is to be developed beyond its present state. This principle will be discussed more fully in the paragraphs that follow.

Strength

To have strength of muscle is to have power. Great power of a muscle, however, does not necessarily mean great size. When a muscle is well trained it has learned (1) to make better use of its fuel supply, (2) to store food fuel in larger quantities, (3) to use increased supply of oxygen and (4) to bring more of its fibers into action.[16] In a person not accustomed to regular vigorous exercise many of the fibers of the large muscle groups are never put to use because the muscle is never overloaded. The more fibers of a muscle that are brought into use, the more power the muscle produces. The importance of developing strong muscles cannot be overemphasized since poor musculature in youth causes health problems in the middle years. However, power of muscles alone is not enough. It is power of muscle plus intelligent use that determines mechanical efficiency.

It is not commonly recognized that the force that can be applied for movement of one part of the body depends largely upon the strength of other parts. For instance strong leg, back, and abdominal muscles influence markedly the ability to catch and hold objects and to throw.

As to the comparative strength of men and women, men are superior. Women as a rule are quite inadequate at exercises requiring suspension or leaning the body weight on the arms. While girls reach their peak in strength at puberty, boys do not attain their peak until late in the twenties.

As we grow older, strength decreases and "as a people we

---

[16]Schneider: op. cit., p. 260.

accept as a normal concomitant of living such multiple physical insults as tobacco, alcohol, and various forms of popular dopes."[17] McCloy states that at thirty-eight years of age the average man has about 85 per cent of the strength of the average eighteen-year-old boy, and women at that same age have only 79 per cent of the strength of the average eighteen-year-old girl. A recent study in a middle western university showed that 33 per cent of the freshman men possessed no more strength than men who were three to seven years older. Had these young men been checked against a standard of fitness for war the record would have been much worse.[17]

When either the intensity or duration of the work is increased beyond the accustomed load, i.e., when the muscle is overloaded, development occurs. Just how much does occur and how soon depends upon the amount of increase in intensity and duration.

Increase in Intensity

Increase in intensity may be brought about by increasing the weight of the load, by increasing the speed, or by increasing the resistance against which the muscle works. Increasing the intensity by increasing the load is shown in the following example: When lying on the back on the floor it is more difficult to lift both legs with the knees straight than with the knees bent, or to lift both legs at once than to lift them alternately. Therefore, if a person does this exercise first by lifting one leg at a time with the knee bent, he "overloads" it when he lifts both legs at once with knees bent, and "overloads" it still more when he lifts the legs with knees straight. The progressions suggested for the strength- and endurance-building exercises in the conditioning series of Chapter IV are based upon this overload principle, and the order suggested should in most cases be followed without change. Increasing the intensity by increasing the speed occurs, for example, when a person has been accustomed to do twenty sit-ups in a certain amount of time and pushes himself to become accustomed to do the twenty sit-ups in a shorter time. Increasing intensity by in-

[17] C. H. McCloy: Militant Physical Fitness and Physical Education. J. Health & Phys. Educ. 14:313 (June) 1943.

creasing resistance occurs when one holds a muscle group in a state of contraction against a heavier-than-usual weight which is tending to break down the contraction, as for example, lifting out at arm's length some object of heavier weight than the muscle is accustomed to and holding it there for the length of time used formerly for the lighter weight. Dr. A. Steindler, Professor of Orthopedic Surgery of the State University of Iowa Hospitals, says that a muscle holding against resistances produces more power than the muscle which is contracting and expanding as it works.[18]

Increase in Duration

There is increase in duration when you increase the accustomed number of times you do an exercise or increase the time for keeping it up. For example, when one usually does Full Squats thirty times at a stretch he overloads and increases the strength of the leg muscles when he increases the number of squats to fifty. Also, when he is accustomed to doing Hip Rolls for five minutes at a time, there is an overloading when he increases the time to eight minutes, or when he can "chin" himself and hold it for a few moments, he overloads by making himself hold against this resistance for a longer time.

Untrained persons should not be given a double dose of overloading by increasing an exercise in both intensity and duration at the same time. If the weight to be lifted is increased, the number of times the exercise is to be done should not be increased at the same time. For example, if a person has been doing fifteen Sit-Ups with arms stretched towards the knees, the dosage is increased in duration by doing it the same way twenty times, or it is increased in intensity by doing the fifteen efforts in less time. The dosage may also be increased in intensity by doing the fifteen Sit-Ups at the accustomed rate but with the hands placed behind the head, thus increasing the weight to be lifted. But if the weight (the resistance) is increased and at the same time the number of repetitions is also increased or the action is speeded up, there is a double dose of overloading and strain may result.

---

[18]What Has Biokinetics to Offer to the Physical Educator. J. Health & Phys. Educ. 13:507 (November) 1942.

Endurance

Endurance means staying power. To develop it an individual needs to develop both muscular strength and cardiorespiratory (heart-lung) capacity and, since endurance develops slowly, he needs to work at it over a long period of time. Since the heart is a muscle, its development also depends upon the overload principle, but it responds to the overload of duration rather than to the overload of intensity.[19]

Steinhaus explains endurance as follows:

> If a muscle is stronger than the load undertaken its motor units function in rotation. This allows portions of the muscle to relax and recover while other units carry the load. If the muscle is just strong enough to match the load all of the motor units must contract at once. Thus increased strength increases the lasting power of a muscle. This is the form of endurance based on strength. It had better be called "local endurance" in contrast to "general endurance" which is due to heart and circulatory capacity.
>
> Muscles can contract and relax without oxygen but they soon fatigue. Oxygen is necessary for recovery from fatigue. If a muscle is supplied with plenty of oxygen it does not become fatigued. Endurance is distance from fatigue.
>
> Since oxygen can come to the muscles only via the blood stream it is obvious that any adjustment that increases the amount of blood going to the exercising muscles will thereby postpone fatigue, i.e., increase endurance.[20]

This general endurance is dependent upon the efficiency of the cardiorespiratory mechanism and is developed by large-muscle activities continued over a long period of time.

The Physiological Versus Psychological Limit In Training

Another important fact to be considered in the development of strength is that there is a limit to the amount of strength which it is possible for a muscle to develop. This limit is set by the inherited number of muscle fibers, but very rarely is this physiological maximum of strength ever attained. Most people stop short of working a muscle long enough to reach its physiological maximum because of their psychological limit which stops them at early signs of fatigue. This limit is reached earlier than is the physiological limit and it is "conditioned by slight aches, respiratory distress, and many mental factors

---

[19] Arthur H. Steinhaus: Fitness and How We May Obtain It. J. Health & Phys. Educ. 14:427 (October) 1943.

[20] A Few Facts and Ideas Presented to the Victory-Through-Fitness Workshop of the National Association for Physical Education of College Women. June 26-27, 1943. (Mimeographed Material.)

such as fear of overdoing, parental warnings, and boredom." It is a safety factor since it prevents overstrain but all too frequently it stops exercise before there has been any overloading and thus there results no development of strength, and proper growth is not stimulated. A person well trained physically has little difference between his physiological and psychological limits. Motivation schemes are useful to reduce this difference. Competition against others or against one's own score or against standards of performance all help to give a person that extra something which it takes to break through the psychological barrier and go on to greater effort towards the physiological maximum of performance for the sake of better growth.[20]

THE MAINTENANCE OF STRENGTH AND ENDURANCE

Once strength and endurance are built up the dosage of exercise should be held at the level arrived at in attaining the goal. Once there is a let-down when a muscle is not exercised at its peak capacity for a period of time the muscle loses power quickly. We speak of that person as having lost strength. This loss can be retrieved only by the original method of overloading. It is much easier to keep in good condition once that condition is attained than to regain it after it has been lost. Research shows that ninety days of inaction will undo sixty days of training.[21] A brief daily, vigorous workout held at the top level of performance already attained will maintain that condition of strength and endurance. It does not require a lot of time. Fifteen to twenty minutes a day will suffice.

THE ROLE OF FATIGUE IN EXERCISE

Although fatigue may result from mental and emotional as well as muscular exertion it will be discussed here only in its relation to muscular activity. In a pamphlet put out by the Army and Navy Department of the Y.M.C.A., Steinhaus defines such fatigue as "loss of use owing to use" and he names the two types of fatigue as the "fatigue of exhaustion" and "the fatigue of depression," the former resulting from long-

---

[21] A. H. Steinhaus: Fitness—A Definition and A Guide to Its Attainment. J. Health & Phys. Educ. 16:175 (April) 1945.

continued muscular efforts which exhaust the fuel supply, the latter resulting from short, intense muscular efforts which clog the muscle with waste products which accumulate faster than they can be removed.[22] Either of these two types of fatigue may be so excessive as to persist after a night's rest. This is known as chronic fatigue or "staleness."

Symptoms of Excessive Fatigue

Symptoms of staleness are as follows:

1. Disturbed sleep
2. Unexplained headaches
3. Loss of appetite
4. Disturbed digestion
5. Constipation
6. Quick exhaustion
7. Irritability
8. Loss of coordination
9. Restlessness
10. Shortness of breath
11. Consciousness of heart beat
12. Poor pulse rate
13. Nervousness
14. Cold feet and hands
15. Loss of muscle tone
16. Muscle tenderness[23]

There is a well-known form of false physical fatigue which occurs even in the absence of hard muscular work. It is merely boredom and it is a psychological rather than a physiological problem. Therefore it will not be discussed here other than to say that usually muscular effort overcomes this false fatigue.

Stages Leading to Fatigue[24]

There are three stages of muscular work leading up to fatigue as follows:

Stage No. 1 is reached by continuous muscular effort lasting for a considerable length of time.

1. Working power gradually increases.
2. Continued performance becomes up-hill work.
3. It becomes difficult to concentrate.
4. A feeling of fatigue becomes apparent and is accompanied by a desire to give up.

Stage No. 2 is reached by the one who persists longer.

1. He gets second wind.
2. Work seems easier now. He is "warmed up" or has found his "stride."

---

[22]Tired?—What Do You Know About Fatigue? International Committee of Young Men's Christian Association, 1944, pp. 3-6.

[23]Arranged from Schneider and Karpovich: op. cit., pp. 235-36.

[24]Ibid., p. 224, arrangement.

3. The output of work remains stationary on a high level.
4. He feels he can continue for a long time.

Stage No. 3 is reached by the one who persists still longer.

1. He comes to a realization that he cannot go on indefinitely after all.
2. The work becomes difficult to do.
3. His muscle power decreases.
4. The work falls off.
5. He is at last fatigued.

How to Ward Off Fatigue

In order to delay fatigue and thus be able to do routine work easily, muscles must be well developed so that they can do considerably more than the run-of-the-day's work. There must be a reserve of energy so that some of the muscle fibers are resting and recovering from fatigue while others are working. This alternating of work and rest among the muscle fibers themselves delays the onset of fatigue in the muscle as a whole. Strength and endurance mean ability to ward off tiredness and thus increase work periods. As Dr. Percy Dawson, the philosopher-physiologist says, "These two qualities [strength and endurance] entitle their possessor to larger slices of everything." These larger slices may refer to mental and emotional activities as well as to muscular effort, because good physical condition delays mental and emotional fatigue as well as muscular fatigue. If, however, an emergency arises which demands more of strength and endurance than has been yet acquired, there are temporary aids for warding off fatigue, such as effective use of muscles as against awkward use (which is energy wasting), and the introduction of rhythm into the work or exercise period. Once one gets "the hang of the thing" which comes from skillful use of muscles, how much easier it is to go on with the activity! Also, how much easier becomes that last mile of a long hike when the group starts a good marching song! Industry is becoming aware of these things and is employing experts to show workers how to get "the hang of the thing" quickly and is introducing music into many of its shops.

The Normal Accompaniments of Activity

Fatigue is not of itself a thing to be avoided. It is a normal accompaniment of activity. Its role is that of protection. When a good night's sleep restores to the previous condition, the fatigue has been normal. It is when a night's sleep does not restore that the fatigue has been excessive and requires attention. However, too much must not be inferred from one night's rest alone for a person's susceptibility to fatigue varies from day to day. As Schneider so wisely says, "The ideal goal is to be sufficiently fit to accomplish each day's work with a minimum of fatigue and to remain active to a good old age . . . with a margin of safety for the experiences of adversity which so frequently descend upon mankind."[25]

[25]Schneider: op. cit., pp. 307-308.

CHAPTER III

THE PLACE OF FUNDAMENTALS OF BODY MECHANICS AND CONDITIONING IN THE PHYSICAL EDUCATION PROGRAM

Since body conditioning plays so important a role in physical fitness, experts in the exercise field should constantly be reviewing their materials and procedures in light of their contribution to the total fitness of the nation. Today the study of the physiology of exercise takes on new importance and points the way to a revision of the physical education curriculum. In the 1920's and 1930's, the emphasis leaned so strongly toward recreation and recreation only,[1] that those who doggedly clung to the tenets of Ling and Jahn found themselves on the defensive with the "moderns." The new program places emphasis on physical fitness for living in a new world where, although recreation is important, it is not the only important feature of a physical education program. Living in the new world calls for a well-conditioned body as fundamental to life's purposes. Indeed, living in the world of any period surely has called for

[1]Dr. C. H. McCloy of the University of Iowa reports in the Journal of Health and Physical Education, June, 1943, that if the city of Des Moines (130,000 population) were to provide enough facilities so that one-half of its adult population between the ages of eighteen and forty-five might engage in daily sports activities it would have to furnish the following:

85 golf links	300 volley ball courts
40 swimming pools	140 track-and-field spaces
300 gymnasiums	5000 tennis courts
300 softball diamonds	

Surely these figures make it obvious that adequate facilities for a rich sports program for any but an imperceptible minority are out of the question in American cities today. Yet many physical educators would use the complete school physical education program to train solely for adult-life participation in activities that obviously can be engaged in but rarely, if at all, and then by only a few.

this, but it has taken the devastating effects of the second World War to make us hear the call once more.

Dr. Walter Truslow, a famous American orthopedic surgeon says:

> As a nation we have lagged behind the totalitarians in body-training of our youth. In the latter part of the last century, more or less adequate physical training had an important place in the programs of our primary and secondary schools . . . Since then school-yard games, rhythmic dancing and more general participation in athletics, valuable as they are, have gradually superceded the scientific posture-training and body-building programs . . . This has resulted in a real let-down in basic preparation for life.[2]

Physical Education Teachers Trained During the Past Decades

The emphasis of the past two decades on the recreational aspects of physical education produced a body of professionally trained teachers who are little skilled in either the practice or theory of body conditioning. The sudden need for vigorous conditioning to meet war emergencies during World War II called attention to the weakness of preparing teachers for a purely recreational curriculum. Of the women physical education teachers trained in the past two decades, few except those well grounded in the so-called "modern dance" have been able to give effective help in conditioning. These teachers have learned the art of body conditioning through the fundamental techniques practiced in modern dance. All other help in the emergency had to come from the few graduates of schools where the "old-fashioned" art of gymnastic teaching had not been abandoned and from the older teachers professionally trained in a day when it was not considered out of step with current educational philosophy to receive thorough drill in gymnastics and the art of gymnastic teaching.

As for men physical education teachers their close association with interscholastic and intercollegiate athletics and their inherent interest in both amateur and professional athletics have kept them familiar with the body-conditioning work of the athletic trainer. It is quite possible that this trainer may not be familiar with the conflicting theories of Gestalt and Bond psychology in relation to the learning of motor skills and the overload principle in the study of the physiology of exercise.

---

[2]Walter Truslow: Body Poise. Baltimore, Williams & Wilkins Co., 1943. Foreword pp. vii-viii.

Beyond that he may not appreciate that there are only a few brave educational philosophers yet among us who have courage to maintain stoutly that hard, disciplined work does have a place in the training of youth for living in a democracy. Yet by some intuitive drive this trainer does know how to teach an athlete short cuts to motor skills, the importance of disciplined training, and the fundamentals for acquiring and maintaining fitness. While the trainer is interested in fitness in order to produce a winner in the athletic arena, the teacher nevertheless must have the same knowledge and interest in the fundamentals of making her pupils fit. Although her hopes may not be quite so tangible as are those of the athletic trainer, they are more important for they are concerned with producing winners for life's arena.

Physical Fitness A Task of the School

A young child who is healthy and robust needs no urging toward participation in vigorous physical activity. However, after years spent in school with its maximum of enforced inactivity and its minimum of vigorous exercise, this natural urge seems gradually to be lost. True enough the schools attempt to forestall this loss by supervised play and by a totally inadequate amount of physical education. The higher we ascend the scale of education the more we find it necessary to require physical education in order that the pupils will secure an adequate amount of participation in vigorous activity.

The physical fitness of a nation has to be planned for and worked for and tackled both early and late in life; and since the schools reach 20 per cent of our population, at a given period, education must do its share in the planning, the working and the tackling.

Most high school pupils enter employment as soon as high school days are over. When a war is in progress they enter employment even earlier. During World War II they worked on varying shifts (day, swing, and night) and they were called upon to do more strenuous work than in peace times. Many young girls combined homemaking with employment and some still in school combined nighttime war work with daytime school work. In the summer of 1945, the U. S. Children's Bureau estimated that one out of three high school pupils was working outside of school and most of them at night work. These growing youth needed protection through bodies that were fit. Indeed, industry demanded that they be fit for they had to stand for long hours (calling for strong feet and correct body mechanics to avoid too early fatigue) and they needed strength and flexibility of arms, back, and legs for lifting, pushing, and pulling, and above all else they needed ability to endure.

The youth of America looks to the schools for its physical training; the schools of America look to the parents for support of the physical education program; the parents of America look to the physical education teacher for help in toughening the moral and physical fiber of their children. This is a task for all who look to American youth to be fit to endure the rigors of world leadership as peacemakers and to be fit to set for the whole world standards of rich, full living.

A GOOD PHYSICAL EDUCATION PROGRAM

A good program in physical education covers a wide range of activities, not only to meet the needs of varying types of pupils but also to meet the varying needs of each individual child. For its share in the development of proper emotional reactions, group consciousness, and sense of personal responsibility within the child, physical education should offer a well-rounded curriculum of sports, games, aquatics, and rhythmic activities; but for certain growth, development, and corrective needs it should also offer conditioning exercises. Some exponents of sports maintain that sports alone furnish all that is needed for growth and development. At best this could be correct only if in every play period each child has a full measure of activity, constant opportunity at the use of sports equipment, frequent turns at those team positions which are most advantageous from an exercise and developmental point of view, and abundant participation in a great variety of sports. However, in far too many sports-education classes pupils spend too much time standing around waiting for turns at equipment or placement on a team and have opportunity at too meager a repertoire of sports to derive full developmental value from the experience. Even were these disadvantages eliminated, too few sports contribute to the development of specific body needs or even to the correction of physical defects to permit sports to hold an all-exclusive place in the physical education curriculum.

As nutrition experts recognize the necessity for a variety of foods to meet various body needs, so do the physical education experts advocate the use of a variety of physical activities in their programs.

A growing child needs meat, vegetables, and sweets in its diet, but it also needs milk, and many children frequently need a corrective diet. In its physical education program it is equally important that the child have not only sports, games, aquatics, and rhythmical activities but also education in the fundamentals of body mechanics and conditioning.

For children who suffer from deformities, malnutrition, and handicapping after-effects of various illnesses, physical education calls for an intensified program in corrective or restrictive procedures under the careful guidance of a physician, but for the normal child such a course in the fundamentals of body mechanics and conditioning as is outlined in this book supplies the "milk" and the corrective diet needed in the physical education program.

A Good Physical Education Diet[3]

A good physical education diet consists of the following:

1. Group team sports, for their socializing influences.

2. Dual and individual sports, for the acquisition of skills to be used not only for the pleasure of the present but for carry-over value for the enrichment of adult life.

3. Rhythms, for their opportunities in personality development and for their contribution to the cultural education of the individual as well as for recreational values.

4. Aquatics, for their training in courage and self reliance, for their importance as safety measures in many life situations and for their carry-over recreational value.

5. Gymnastics, for their particular value in body conditioning, body development, and the correction of postural defects.

6. Fundamentals of body mechanics, for their special value in teaching the skills of body management in everyday tasks.

[3]The activities listed here contribute to general growth and developmental needs only when engaged in strenuously enough to produce real physiological stimulation.

THE NEED FOR A COURSE IN THE FUNDAMENTALS OF BODY MECHANICS AND CONDITIONING

Every child should be given exact and definite information about the exercise needs of the body and beyond that should be given abundant opportunities for the practice of such bodily movements as will give sufficient strength, endurance, flexibility, relaxation, agility, and kinesthetic awareness to meet the ordinary demands of life. Still further, he should be given knowledge of and practice in the skills of body control so that he can handle his body effectively for ease in the accomplishment of tasks, for satisfaction in social relationships, and for enjoyment of recreational activities. Such educational material and practice calls for a special course within the physical education program, a course which should be the keystone of the physical education of the child. Added to this there should be training in those skills that are fundamental to sports, aquatics, and rhythmics; this in turn to be followed by vigorous and frequent participation in the activities themselves. However, like a major theme in a symphonic poem, training in the fundamentals of body control should ever be a recurring part of the physical education program at all age levels. Just as food must be taken day by day for constant refreshment, so must conditioning be a constant in the attainment and maintenance of physical fitness.

The authors do not claim that a course in the fundamentals of body mechanics and conditioning is the only means of supplying "milk" for the physical education diet. Superior opportunities in sports, aquatics, or rhythmics may be a good substitute, but it is only an infinitesimally small percentage of people who are fortunate enough to have at their disposal the proper equipment and facilities or sufficient time, under the tutelage of sufficiently competent teachers or coaches, to derive the necessary benefits. For the great rank and file of people, conditioning and the fundamentals of body mechanics must be acquired under circumstances that require but little and inexpensive equipment and facilities and call for the type of activity that can be conducted by relatively inexperienced and untrained teachers. Even under circumstances favorable for one to be physically educated through sports, rhythmics, or aquatics alone, the authors wish to call attention to the fact that the best dancing, sports, and swimming teachers and coaches use so-called "conditioning exercises" or "fundamental techniques" to supplement the dancing, sports, and swimming practice.

The Objectives of a Fundamentals Course

The objectives of a course in the fundamentals of body mechanics and conditioning may be grouped under the following headings: physical, psychological, educational, social, and economic.

Physical Objectives

The physical objectives of conditioning exercises are strength, endurance, coordination, flexibility, relaxation, agility, and kinesthetic awareness. These will be discussed later in Chapters IV and VII.

Psychological Objectives

Fitness is achieved and maintained only through disciplined effort. Discipline, whether self-imposed or enforced by others (yet accepted in good spirit), cannot fail to produce good effects on one's spiritual development. Difficult tasks, tackled and surmounted, are important building stones for the sort of spirit that has made America what it is, the only spirit that will carry America safely into world leadership of tomorrow. It is not soft, undisciplined people who will control the peace. Nothing worth-while is ever achieved by the undisciplined. The fear which so many educators show toward this whole problem of disciplined education of school children is one of the greatest faults of today's educational system. Fortunately there are still left in our schools many educators who, though labelled "old fashioned," staunchly stand by their guns and surround the children under their care with opportunities both to practice self-discipline and to accept discipline from others who have been accepted as leaders. This is the sort of education upon which democracy is grounded. Physical education makes a real contribution in this direction through the self-imposed tasks which it offers to children in their efforts to train the body. There is positive morale training in a fitness program conscientiously pursued.

Educational Objectives

For intelligent learning the child must understand not only the what, but the why, who, where, how, and when of his school subjects. The educational objectives of a course in the fundamentals of body mechanics and conditioning are just these, namely, to give the child an understanding of what fitness is, why it is necessary, how it can be acquired, when and where it is to be pursued and by whom. These objectives are reached by a definite dosage of information accompanying the practice of the activities. Such a dosage is explained in the chapters that follow.

Social Objectives

For ready social acceptance it is of prime importance for an individual to have good physical bearing and poise. It also adds to one's "staying powers" socially to be skilled in a variety of recreational activities so that one can be a good companion for leisure hours. Being fit, feeling "tops," having a body well trained in the fundamentals of body mechanics, which of course includes training in good carriage, are important features of a passport to social acceptance. Beyond this it is important to be accepted not only as a desirable member of one's social group but also as a worth-while citizen of one's community. If the course is conducted so as to meet full social objectives it will give the pupil an opportunity to evaluate herself in relation to her group, to arrive at wholesome attitudes and appreciations of her own weaknesses and strengths, to acquire an understanding of her body as a healthful organism and of her potentialities as a member of a democratic society.

Economic Objectives

If it is important socially to have a good carriage and to be well poised, it is even more important for success in one's life work to have these desirable qualities. To look well and to feel well is a long step towards the ability to work well and to get on well with one's fellow man.

Conditioning—The Keystone of a Fundamentals Course

Just as a course in the fundamentals of body mechanics and conditioning is the keystone of a physical education program, so conditioning exercises are the keystone of this fundamentals course.

Conditioning exercises carefully planned and vigorously pursued will build strength, flexibility, and endurance. Upon these the child must be taught to superimpose the skills of body control and body management which will give him efficiency for both the tasks of everyday life and the pleasures of everyday living.

Body Management Skills

This fundamentals course should cover the skills of posture (both standing and sitting and involving the use of the body in varying situations) and such life-situation activities as pushing, pulling, stooping, lifting, and carrying objects of varying weights and bulks, climbing stairs and ladders, throwing, catching, and striking objects of various sizes, and dodging obstacles both moving and stationary. Beyond this, the course also should give instruction incident to the variety of pursuits to be followed by various class members. For example, in emergency periods when many girls and women expect to replace men in factories they need to be thoroughly grounded in the skills of climbing high ladders and walking catwalks with one arm full of equipment. If some are to do welding they particularly need quickly to develop strength of arms, shoulders, and upper back in order to hold up equipment for long stretches of time in overhead positions without undue fatigue. They also need freedom of movement in the thigh joint and strength of sacro-iliac joints to support the frequent one-knee kneeling position which gives quick backache to the weakling. Also, they must do much climbing, both deep down and high up. Those girls who are to work on farms should be taught how to use the body most effectively when using a hoe and rake, picking garden truck, lifting heavy feed bags, spading, cleaning pens, closing heavy barn doors, cranking tractors, and doing many other such farm chores. What American girl does not expect to do her share at home in hanging out clothes, carrying heavy clothes baskets, ironing, moving furniture about, closing garage doors, helping with gardening and lawn work and doing countless other physical tasks—all tasks that can be done the wrong way, causing frustration and unnecessary weariness or the right way, giving satisfaction and building still more strength and endurance for tomorrow.

Conditioning and body management skills make up the subject matter of such a fundamentals course. There is an infinite variety of material to draw from in constructing the course. Chapters IV, V, and VI present material designed to meet the conditioning needs of the general run of school girls while Chapter VII presents material for the study of and practice in body mechanics.

Those skills that contribute to the effective body management in the pleasures of everyday living, such as skills in handling sports equipment and skills fundamental to good swimming and dancing, are usually taught in physical education classes dealing directly with those activities so that they are not as a rule included in a fundamentals course; but the skills of body management in the tasks of everyday life are the neglected skills unless incorporated into a special course.

THE NEED FOR CONDITIONING EXERCISES

Any exercise that will help attain and maintain that part of physical fitness concerned with strength, flexibility, and endurance is a conditioning exercise. The pursuit of these objectives is the only logical reason for using gymnastics in the educational program; however, this alone is reason enough. No educator need apologize for including gymnastics in his program even though the too ardent advocates of educating through sports alone view with a scornful eye any one who has the temerity to speak in defense of gymnastics as one of the necessary tools in the physical educator's kit.

There are probably as many conditioning exercises as there are people versed in the study of kinesiology and physiology of exercise to devise them. Any exercise that will put any part of the body into good condition is a conditioning exercise. There are literally hundreds of conditioning exercises from which to choose a set. By a set is meant a series of exercises which taken as a complete unit will condition the body as a whole. To be good, a set should contain exercises for the neck, arms, shoulders, upper back, lower back, abdomen, trunk as a whole, legs and feet, and, if it is to include general endurance building as well as strength and endurance building of various parts, it should include some exercise that gives the entire body enough of a workout to cause deepened breathing and quickened heart action.

All Activities Not Equally Good as Conditioners

Just as one kind of food meets many needs and may overlap somewhat the needs met by other kinds of food, so the various activities of a well-rounded physical education program overlap in their objectives at many points. While some group, dual, and individual sports may be excellent body conditioners, many sports ordinarily engaged in do not meet this objective. Also it is quite possible for a few pupils to get excellent physical developmental work out of a sports class while the great majority of the members of the same class get but little organic stimulation, owing to insufficient equipment or inadequate opportunity to take their turn or to play in positions offering strenuous activity.

Similarly, rhythms and aquatics may be excellent body conditioners, but as usually taught by the unskilled teacher it is so easy to fall short in this direction while admirably fulfilling other objectives.[4]

Gymnastics the Ideal Conditioning Activity

To assure that body conditioning and body developmental work is received by all pupils and at an adequate dosage it is imperative that, in addition to sports, a form of activity be taught in which constant mass participation can be secured and in which individual proper dosage of exercise can be administered. Such a form of activity is found in gymnastics[5]

---

[4]One exception to this statement is found in that form of rhythmical activity known in physical education circles as the "modern dance" which is not to be confused with the modern dance of the ballroom variety. This is a distinct type of dance which has replaced the so-called "aesthetic dancing" of earlier days. In its fundamental techniques it offers vigorous body conditioning equal to anything offered as gymnastic exercises. In fact these techniques are of gymnastic form done to music. A good teacher of modern dance gives her pupils as excellent body-conditioning work as does a good gymnastic teacher. The one can substitute for the other with no loss of opportunity to meet objectives, and the modern dance teacher has the advantage of going on from there to the creative, personality, and cultural development of her pupils.

[5]Using the term in its usual connotation as covering free-standing exercises and exercises done with either light or heavy apparatus.

or in the gymnastic form of rhythmical training such as is found in modern dance.[6] In this form of activity large groups can be handled by one teacher and without equipment— a situation all too frequently forced upon the physical education program.

Since human beings are more alike than different, the great rank and file of children of any given age group have much in common—the same growth and developmental needs and the same body mechanics needs for everyday living. Therefore it is not ignoring the individual needs of the child to put the great mass of children through a common experience in the education of their bodies towards a richer and fuller life.

The Use of Gymnastics Should Not be Abused

If conditioning through gymnastic exercise is to be offered in the school program it should be there for the sake of its own objectives and should be taught intelligently. It should not be used merely as an excuse for not offering a rich sports program because of lack of facilities, equipment, and properly trained teachers. True enough, although never a teacher saver, it is a space saver, a time saver, and a money saver, and too many school administrators use it as a crutch to hold up their utterly inadequate plans for an all-round educational program in physical activity. Because of this universal abuse over a period of many years, gymnastics has come to be looked upon as an inadequate substitute for better things. It has fallen into disrepute, and sight has been lost of the fact that it does have merit on its own and that it can be used educationally to meet important objectives as no other activity can meet them—this of course provided the teacher is adequate. But isn't this a necessary corollary to all teaching? Why must we always be stressing the fact that the educational objectives of physical activities are not fulfilled without the leadership of adequate teachers? No doubt the answer lies in the fact that the great majority of pupils in American schools today receive their

[6]The full objectives of the acquisition of proper strength, flexibility, and endurance of the entire range of large muscle groups are not met with, however, in the techniques of modern dance unless the teacher understands what is needed for development and conditioning and puts into her fundamentals work the necessary exercises to meet these needs. But the same thing may be true of gymnastics, also, if the teacher permits this objective to be clouded by other aims.

physical education from teachers who are woefully unskilled in the art of teaching this subject. Even the few so-called "trained teachers" are all too frequently trained too superficially to be adequate. Too many teachers merely exercise their pupils, exercise them without plan, without purpose. And too many others do not even exercise them, permitting the class hour to be filled with aimless standing around, waiting "turns" at the little equipment available.

Conditioning a Continuing Need

Conditioning exercises are not to be looked upon just as a foundation which once laid is forever finished. Instead they are like the musician's scales which, although fundamental, are forever after a "daily must" if one is to keep in proper training. It is said that Fritz Kreisler, famous violinist that he is, practices the scales every day. No doubt this is true of all good musicians. Conditioning exercises are the "scales" of the physical education program. Since they are fundamental not only in attaining fitness but also fundamental in the day-by-day maintenance of fitness, it is urged, therefore, that teachers of physical education not only use conditioning exercises as the foundation for their fundamentals course, but that they also inspire their pupils to do the exercises at home each day. Beyond this, teachers should use conditioners as a "warmer-upper" for every sports and aquatics class. Good athletic and swimming coaches use such exercises to start off every practice period and every athletic contest.

The Place of Conditioning Exercises in a Program of Living

The conditioning exercises offered in this book are designed for use first, in the school, and secondly, in the home.

The way in which they are taught and motivated in the school will determine to what extent they will be used in the home. If the same series is used throughout all age levels the various school members of the family will be familiar with the same set of exercises and thus may derive enjoyment and profit from doing them together at home with the parents, brothers, and sisters drawn into participation.[7]

---

[7]For excellent suggestions for such family participation see Frederika Marston: Be the Gym Teacher in Your Home. J. Health & Phys. Educ. 14:14 (January) 1943.

If one series of exercises is chosen and then strictly adhered to for a sufficient length of time to insure their becoming as familiar as washing the face and combing the hair, they should become useful as that tool of education which individuals can use in after-school years to help maintain good condition. The tempo of present-day American life almost precludes daily participation in an active recreational sport for a long enough time to get any real stimulation, to say nothing of the additional amount of time necessary in travelling to and from the golf links, the tennis courts, or the swimming pool. Why not recognize this as an impossibility for most adult Americans as a daily practice and urge the use of ten or fifteen minutes of conditioning exercises instead? If such a series of exercises has been thoroughly learned at school, their adoption for a daily short exercise period at home becomes much simpler.

This does not mean it is desirable to substitute conditioning exercises for sports and recreational activities. It is merely a recognition of the fact that one's exercise diet should consist of both the main course and the dessert. However, it does mean a reversal of the usual emphasis upon the place of each in a program of living. Instead of urging sports participation every day with conditioning exercises substituted whenever the former is unattainable, it would seem more logical to insist upon the conditioning exercises as the daily procedure with participation in vigorous recreational activities added at every possible opportunity.

If, in addition to teaching the correct execution of each conditioning exercise, the teacher gives the pupils information about the main purpose of each movement and the region of the body affected, each individual will have a working knowledge not only of how to keep himself in good physical condition, but of how to combat or offset the physical strains and drains which may be peculiar to his day's work.

In a school where the above has been put into practice students frequently volunteer the interesting information that they have tried several of the flexibility exercises on evenings when their necks and shoulders were tired and aching at the close of study periods and that the results were such relief and relaxation that they are doing the exercises every night. In other instances girls who complained frequently of tired feet have made the association between this condition and the remedy and are observed doing outside of class a foot exercise of the conditioning series which they have been taught in class.

Legion are the school girls who, having in physical education class learned the correct skill of "abdominal pumping" and acquired a thorough understanding of its purpose, have passed on this skill and knowledge to sisters, aunts, and mothers, thus earning their deep gratitude because of the release which the exercise brings from menstrual period discomforts. People who have undergone a recent operation are frequently advised by their physicians to start taking conditioning exercises as soon as they leave the hospital. Fortunate indeed are those patients who learned such exercises in school and did them so frequently that they still remember them in this hour of need. More than ever after World War II physicians are prescribing conditioning exercises to their patients, having witnessed in the Army and Navy hospitals the greatly curtailed convalescent periods from an early start at conditioning even while the patient was still confined to bed. Conditioning exercises thoroughly and correctly learned in school days will stand one in good stead in an emergency. In fact a daily workout at conditioning exercises followed conscientiously week by week, month by month, year by year, may well forestall many emergencies.

Such incidents as those stated above lead one to assume that more than average familiarity with a certain set of exercises, if combined with a knowledge of what they can do for one, cannot but help foster their continued use. With these advantages in mind the authors have carefully selected a definite group of exercises to make up one series as described fully in Chapter IV and they suggest the use of this one unless the teacher, sure of her kinesiological analysis, cares to make substitution from the dozens of exercises which are equally good.

Formal education is useful only as it contributes to the growth of the individual and to an understanding and appreciation of the place and use of the various tools of education in building a program of living. Therefore it is just as important that students learn the place and use of conditioning exercises outside the school as that they learn to do them accurately and thoroughly in the school.

CHAPTER IV

CONDITIONING EXERCISES

Within the last few decades the pendulum of interest in a gymnastic type of exercise has swung from the one extreme of too much emphasis to the opposite extreme of complete repudiation. The pendulum is now swinging toward the midway point. Perhaps experience of the two extremes was necessary before gymnastic exercise could fit into what now seems to be its proper place in the whole physical education program. Its objectives are specific but by no means all inclusive. Emphasis should also be placed upon recreative skills for leisure time and upon other group and individual experiences which will contribute toward democratic living.

The exercises presented in this chapter are designed to help individuals to attain certain definite objectives conducive to the development and efficient use of the body and the maintenance of physical fitness in so far as exercise contributes to total fitness. The intention was to provide a body of educational material which could be mastered, understood, and remembered well enough to fill a need in the future as well as in the present. The important factors which control the success with which these goals are realized are discussed briefly.

Each exercise is described in detail as to body position and movement. Suggestions are made of changes which will progressively increase the difficulty of the movements. The purpose of each exercise is given in terms of specific objectives. A kinesiological analysis of each exercise has been made to make certain that all important muscle groups are included in the series. The teamwork among muscles of the body is so intricate that it is almost impossible to single out a particular muscle or muscle group to be used without involving others (Figs. 1 and 2). Each exercise presented, therefore, was chosen for its value in developing certain parts of the body, and in the statement of purpose any concomitant results have for the most part been purposely ignored.

Certain important cautions which should be heeded for the correct execution of the movements have been added to the description of each exercise. Further teaching suggestions particularly designed to aid the inexperienced or untrained teacher may be found at the end of this chapter and in Chapter IX.

OBJECTIVES

Conditioning exercises should provide opportunity for the development of those specifics of strength, endurance, coordination, flexibility, relaxation, agility, and kinesthetic awareness which are basic to the ability to use the body efficiently for whatever task life imposes. For those individuals who, in after-school years, are too busy to participate in active sports they provide a short-cut exercise program for the maintenance of physical fitness. To fulfill this purpose a small number of exercises should be learned very well rather than a great number sketchily, and pupils should be taught as well as exercised!

Strength and Endurance

Strength and endurance have been discussed in detail in Chapter II and therefore will not be elaborated upon here.

Coordination

A person may have strength and endurance and yet lack skill. Rarely, if ever, does one muscle act alone to produce a movement. It is the way in which several muscles act together which produces the skillful performance. Some of these coordinations are inherited and instinctive, such as winking the eye, coughing, or sneezing, while others must be learned. This learning process comes easily to some people and with difficulty to others, but skill can be achieved to a certain degree even by those with poor motor ability. A coordinated act involves the nerve impulses that order the action, the selection of the particular muscles to act, and the amount and sequence of their acting. It is developed through training and practice. Most people have no idea what muscles they are using from

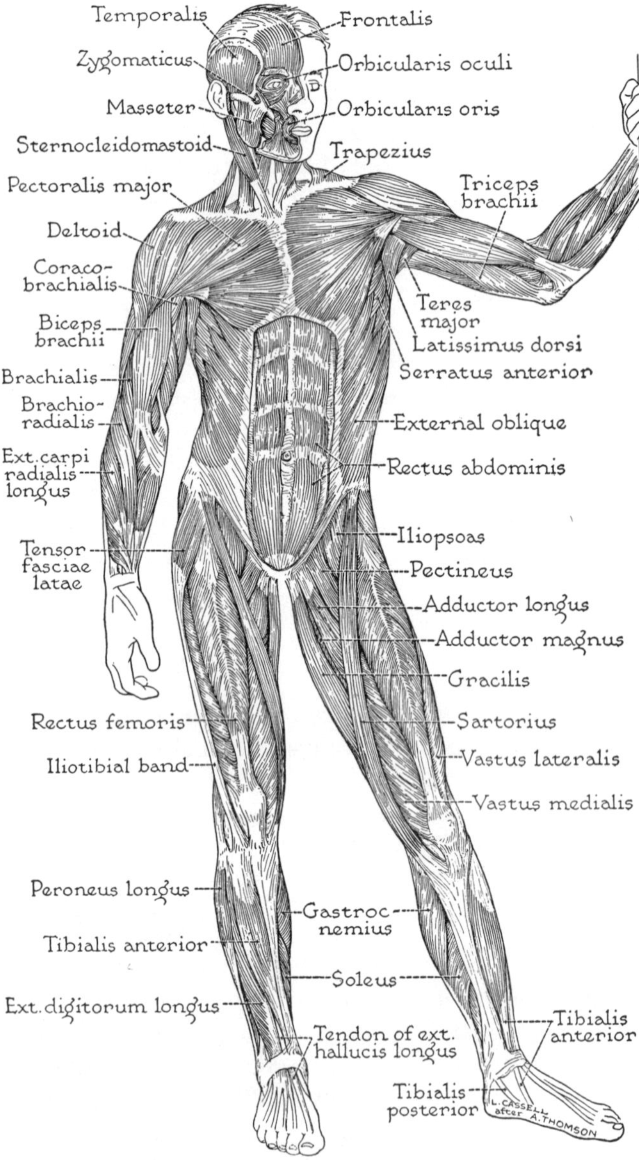

Fig. 1. Anterior view of body muscles.
(Millard and King: Anatomy and Physiology.)

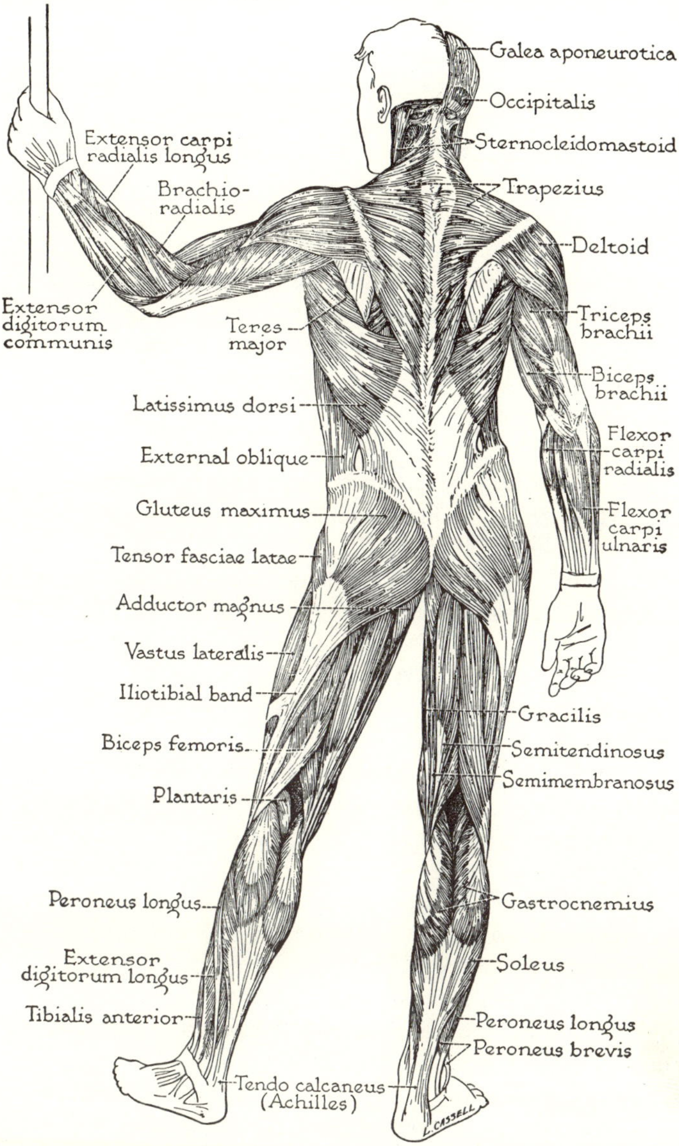

Fig. 2. Posterior view of body muscles. Note the biceps femoris, semi tendinosus, and semimembranous which make up the group commonly called the hamstring muscles. Also note the lower end of the rhomboid muscle just barely showing under the lower edge of the trapezius and upper edge of the latissimus dorsi and to the right of designated teres major. It lies under the middle part of the trapezius and runs from the scapula diagonally upward to attachment on the seventh cervical and first five dorsal vertebrae. (Millard and King: Anatomy and Physiology).

any knowledge of the muscles as such, but they do recognize that certain actions produce satisfying results. These are the movements they tend to repeat. This "trial-and-error" learning process can be speeded up considerably by criticism and suggestion from a good teacher (one who can analyze the activity and break it down into its fundamentals) and it can be speeded up still more by skillful demonstration.

A coordinated act is also a more efficient act. As coordination increases, the amount of energy used in performing the movement decreases.[1] Therefore, muscular movements that are efficient are also useful, economical, and graceful.

Flexibility

To develop flexibility means to increase joint mobility, or to increase the range of those motions which are possible in a joint. Exactly what happens in the tissues themselves is unknown but flexibility is accomplished by stretching or elongating tense and overshortened muscles. Lack of tension in those muscles to be stretched is of foremost importance, and to assure this it is necessary that flexibility movements should not be forced, especially with beginners, but should utilize the momentum developed in free and rhythmical action. As the necessary relaxation is acquired, more and more vigorous momentum may be used, and since the movements are all self induced this can be regulated by each individual to her own satisfaction.

The use of momentum and the weight of body parts in free and rhythmical movements may not be as efficient a corrective procedure for stretching muscles as some other type of movement.[2] The exercises presented in this chapter, however, are designed primarily for attaining and maintaining body conditioning and not for correction in its usual connotation. The pupil who has marked postural deviations calling for individual corrective procedures should be in a special corrective class.

---

[1] Percy Dawson: The Physiology of Physical Education. Baltimore, Williams & Wilkins Co., 1935, p. 215.

[2] Too little is known at present about how the action of the stretch reflex and reciprocal innervation affects the stretching of muscles. More information at some future date may help us in solving this problem.

The degree of suppleness attained by acrobatic performers is not necessary nor even desirable for the average individual. However, a condition of stiffness of the muscles and ligaments in the shoulder girdle, chest, hip joints, spine, and ankle joints may be factors contributing to poor posture and inefficiency as much as weak musculature. Such a condition needs correction.

Relaxation

Relaxation is discussed in Chapter VII. The reader is referred to this material instead of duplicating it here.

Agility

The development of agility is dependent upon the development of sufficient flexibility and strength to perform the desired movements and the coordination to do them skillfully, lightly, and easily. It may be developed through participation in activities which demand a quick type of motor response, such as Squat Jumps, rope jumping, and Squat Thrusts. Vaulting over low obstacles, stunts and tumbling, and games demanding quick changes in direction, such as chasing games or basketball, are other excellent activities for the development of agility.

Kinesthetic Awareness

To develop kinesthetic awareness means to develop an accurate and conscious control of the body in movement. When a muscle is contracted certain sensations arise in the muscle and the joints involved in the resulting movement by which an individual becomes aware of what that movement "feels" like. He learns the difference between tension and lack of tension, not through any external means such as observation with the eyes, but through the sensations arising in the muscles and joints themselves. He can tell with closed eyes whether he is moving forward or backward, whether the arm is straight or bent, whether it is held at the side or out at shoulder level. This is kinesthetic perception, and if it is to be relied upon as an accurate tool in learning physical skills, movements should be done with conscious accuracy during the learning process, which means insistence on good form in performance. At first the eye of the performer himself, of a partner, or of the teacher must check the accuracy of this kinesthetic awareness until it has been developed to that degree where the performer knows with confidence "from the feel" that the body or any of its parts is exactly in the desired position or is moving in the desired movement.

To aid pupils in acquiring kinesthetic awareness a teacher should be careful that all exercises are accurately demonstrated and that the first attempts of the class are carefully controlled and criticised so that the movements are learned correctly at the very start. Once done in a certain way the tendency is to repeat the performance in the same manner whether correct or incorrect unless changed through conscious effort. If the performance is accurate a comment to that effect gives the student confidence in its repetition, but if it is incorrect, criticism should be given before the movement is repeated. Suggestions should be made to improve performance if kinesthetic awareness is to be developed accurately.

A SUGGESTED SET OF CONDITIONING EXERCISES

Two Schools of Thought

There are two schools of thought about conditioning exercises. One school made up of the idealists contends that the teacher should teach a great variety of conditioning exercises to pupils in the hopes that the variety will interest them; that some way, somehow, they will remember a lot of them, will remember how to do them properly, will not get confused on the conditioning purpose of each, and in after-school days, will draw from this wide fund of knowledge to give themselves conditioning as needed. The other school of thought is made up of the realists. They believe that if a pupil explores too wide a field he learns nothing "for keeps." The followers of this school believe that the teacher should select a good set of general conditioners for the average pupil and that the pupils should be taught this one set and this one set alone so that they will thoroughly master the skills of each exercise in order to receive full benefits trom it; that they will learn the purpose of each exercise so that they will know how to use it wisely; and that they will acquire a full understanding of the steps of progression of each exercise in order to build strength and endurance.[3] Such a set of exercises is "the scales" upon which the teacher starts to build a "musical career." Of course it may get a bit boring; so does practicing scales become boring, but what musician worthy of note ever denied the scales because they bored him? It takes constancy of purpose to become a musician; it also takes constancy of purpose to keep the body in good condition.

Once the pupils have been drilled and drilled in one set of conditioning exercises until they have mastered it so thoroughly that they will remember it all the rest of their lives they have something to build on, something sure, something stable in times of stress.

Common Use of One Set Throughout the Program

Once body movements are analyzed into their component elements it is obvious that there are certain movements fundamental to all activities which may be developed through the medium of gymnastics, dancing, sports, or aquatics. Therefore it is both logical and correct to use the same series of conditioning exercises in all of these activities. In classes in sports, aquatics, or fundamentals of body mechanics and conditioning the voice may be used to set the rhythm, while in dancing or rhythm classes tom-tom, piano, music box, or other musical instrument may be used, although the type of exercise in each case may be the same.

If one good set of conditioning exercises is selected and the same one used in every physical education class period in one school unit, it may foster more general participation outside the school just as a common language pattern fosters better communication.

The average individual does a great many daily tasks without particular effort, such as dressing, brushing the teeth, and

---

[3]The Women's Commission of the Committee on Physical Fitness of the Federal Security Agency went on record at a meeting in Washington, D. C., March 1945, favoring the teaching of at least one exercise series to all adolescent girls so that they will know one set of conditioners to use when other activities are not available for daily participation.

combing the hair. With the child there is first a stage of learning how to do things correctly, then comes motivation by the parent to do them at certain stated and necessary times of the day until, finally, through constant repetition and great familiarity, the adult does them with little effort or thought. So it is with conditioning exercises.

Devising a Set of Conditioning Exercises

For "the scales" the teacher should select most carefully the exercises that will make up her set. She must be sure of the kinesiological analysis of each exercise so that it will for a certainty do the developmental job it is supposed to do. Daily newspapers and current women's magazines frequently offer to their readers exercises that will never in the world fulfill their supposed functions, exercises most obviously devised by persons untrained in the science of body development. Perhaps the greatest of these crimes is committed in the name of "reducing exercises." Yet, on the other hand, we frequently find excellent and correct exercise suggestions in these same newspapers and magazines. Let the teacher be on her guard in devising the set of conditioning exercises which she will give to her pupils. For the guidance of the untrained or the inexperienced, the authors present in this chapter a set of conditioning exercises that has been analyzed kinesiologically and which will, if executed according to the full instructions accompanying them, fulfill the purposes as stated. This set has been given a thorough tryout by many groups of girls who have been tested frequently and who have stayed with this one set exclusively for a long period of time, long enough to give trustworthy testimony concerning its effectiveness for their needs. As a result of this tryout and criticism the authors present this set for those who prefer not to devise their own set. It is not the best set. Who knows what is the best set? It is a set and a good one! Any one else professionally trained and experienced can devise an equally good one. If it makes the path of any physical education teacher a bit easier by using this set instead of going to all the time and thought it takes to originate her own set, the authors will feel rewarded for their time and thought.

A Conditioning Series That Has Been Successfully Tried Out

Name of the Exercise	Main Purpose of the Exercise
1. Warmer-Upper	1. To stimulate the circulation, for warming-up, and to increase cardiorespiratory endurance.
2. Arm Circling	2. To reduce muscle tension and increase flexibility of shoulder girdle.
3. Body Swing	3. To reduce muscle tension.
4. Trunk Twister	4. To increase flexibility of the shoulder girdle, spine, and thorax.
5. Trunk Limbering	5. To increase flexibility of spine and legs.
6. Full Squats or Squat Jumps	6. Foot, leg, and thigh strength.
7. Sit-Ups	7. Abdominal strength.
8. Hip Raising	8. Lower back and hip strength.
9. Hip Roll	9. Abdominal strength.
10. Push-Ups	10. Arm and shoulder girdle strength.
11. Wing Lifts	11. Neck and upper back strength.
12. Abdominal Pumping	12. Increase pelvic circulation and relieve dysmenorrhea.
13. Foot Rocker	13. Strengthen muscles supporting the longitudinal arch of the foot.

DESCRIPTION OF THE EXERCISES IN THE SERIES

The full description of each exercise, including the cautions, should be read carefully before attempting the movements. After trying the movements the description should be read again to be certain the exercise is accurately interpreted.

EXERCISE 1. WARMER-UPPER (RUNNING OR ROPE JUMPING).

Purpose

To stimulate the circulation, for warming-up, and to increase cardiorespiratory endurance.

Execution

See Chapter VI for detailed description of proper use of the body in running and the technique of rope jumping with variations which may be used.

Progression or Variations

1. In classes on the gymnasium floor it is suggested that the run be timed rather than made a certain number of laps

around the floor. Call the time aloud at thirty-second intervals, such as "thirty seconds . . . one minute . . . one minute, thirty seconds," and so on. In this way each girl can check on her own progressive ability to run for a longer time. Be sure the whole group understands that they are to run as long as they can without too much distress then, when necessary, drop to the outside of the line and continue by walking. As soon as possible each individual should get back in the line and run again. The length of the initial run before dropping out to walk should gradually be increased, and by calling the time aloud each girl can check on her own ability to do this. Start with one minute and a half and increase by fifteen-second intervals as rapidly as the group seems able to take it until three or four minutes is reached, then maintain it there.

2. If done in the home or in a room where there is insufficient space for ordinary running, running in place with high knee lifting may be substituted. Here again, note the time at thirty-second intervals and mark time in place, swinging the arms easily back and forth when it becomes necessary to rest a bit. As soon as possible, start running in place again.

3. If preferred, rope jumping may be substituted for running, but if there is to be any increase in cardiorespiratory endurance there should be the same motivation to continue the jumping for as long a time as possible as there is in the running.

4. A run around the block, or a certain number of blocks may be used by an individual at home as a Warmer-Upper for the conditioning exercises to be done, or by an organized class. It is suggested here that the individual time herself on running around the block and stimulate herself to shortening this time. When this distance can be run easily at a fast pace, lengthen the distance and repeat the timing process.

Cautions

1. It may be wiser to omit rope jumping during the first two days of the menstrual period and possibly the last two days preceding the onset of the flow. Running may be substituted for jumping at these times.

2. Rope jumping should be light and easy, and just high enough to clear the rope. All redundant motions of the feet

and legs should be eliminated, such as excessive knee action resulting in a kicking motion.

3. Endurance running should be done, of course, only by the individual who knows for certain that her heart and lungs are organically and functionally sound.

4. After running it is important to continue with walking, marking time in place, or some other light arm or leg movement, or both, because such movement favors recovery. Complete relaxation is not the best way to recover after strenuous exertion.

5. Always emphasize running in good form.

EXERCISE 2. ARM CIRCLING.

Purpose

To reduce muscle tension and increase the flexibility of the shoulder girdle.

Execution

1. Position. Stand erect, feet parallel and 3 to 4 inches apart, arms at sides.

2. Movement. On count one, step forward with the right foot, place the right hand on the hip, close the left hand into a tightly clenched fist, and start the left arm in a circle, forward-

Fig. 3. Arm Circling.

upward. Continue swinging the left arm forward-upward, then backward-downward so it makes a complete circle (Fig. 3), until seven circles have been completed. On count eight, bring the right foot back beside the left foot, place the left hand on the hip, and lower the right hand from the hip to hang by the side. Repeat the circling with the right arm, clenching the fist, and stepping forward with the left foot on the first count. On count eight, bring the left foot back to the right, place the right hand on the hip, and lower the left arm to the side. Repeat once more on each side. Total 16 circles with each arm.

3. Setting the Rhythm. Establish the rhythm by the use of counts. It may help at first to substitute the word "change" for the eighth count as, "one-two-three [etc.] seven-change" then repeat "one-two-three [etc.] seven-change," denoting the change of arms and the reversing of the foot position.

Cautions

1. The shoulders should be kept squarely to the front and not allowed to turn toward the circling arm.
2. The circling arm should be kept straight throughout and be made to feel heavy and long.
3. The fist should be clenched to keep the blood from being forced into the fingers resulting in an uncomfortable sensation.
4. The rhythm should be quite slow and deliberate so that the full range of circling movement may be obtained. **Never speed up the rhythm.**
5. The arm should make as large a circle as possible, almost brushing the ear when overhead, and going as far backward as possible on the downward movement.
6. The arm should be swung forcibly so that the fist describes an ever widening circle on each swing.
7. Demonstrate, presenting a front view to the class.

Progression

Progression in this exercise will be noticeable first by the manner in which the movements are executed, particularly the increased relaxation of the arm and shoulder and the resulting increased range of the circling movement. There will also be a gradually increasing skill in the smoothness of the transition from one side to the other, in the change of both arm and foot positions.

EXERCISE 3. BODY SWING.

Purpose

To reduce muscle tension.

Execution

1. Position. Stand erect, feet in a forward-backward stride, toes pointing straight ahead, arms raised overhead in a relaxed manner (not completely stretched nor vigorously extended upward).

2. Movement. The arms, head, and trunk are allowed to drop forward-downward in a completely limp manner, the

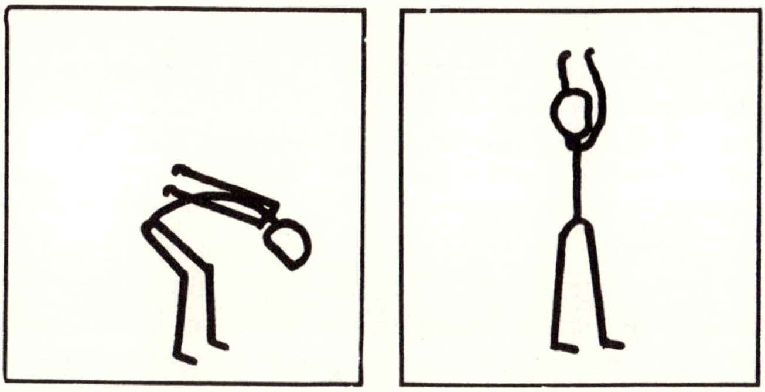

A B

Fig. 4. Body Swing.

trunk going down to approximately a horizontal position (Fig. 4, A). The knees are bent slightly. The momentum of the relaxed arms carries them downward and on backward past the thighs and up in the back. At the end of the backward-upward swing of the arms, when momentum has been used up and the natural recoil begins, the back, leg, and shoulder muscles gradually take hold and lift the body to the erect position again with arms overhead (Fig. 4, B). The whole movement should give the appearance of a limp, heavy, pendulum swing with no jerky stops or pauses anywhere. Repeat eight to sixteen times.

3. Setting the Rhythm. In order to secure release of tension the flow of movement from flexion to extension and back to flexion again must be done with no held positions nor static body parts. Thus each individual should establish her own rhythmic pattern and there should be no attempt to keep the group together unless it is done to music. In this case the movements should follow the rhythm of the music.

Cautions

1. Watch for the following indications that relaxation is not secured:

a. Straight and extended fingers—if relaxed the fingers will be slightly curled.

b. Stiffly extended elbows.

c. A jerky knee bending and stretching.

d. Stopping the downward arm swing suddenly just at the thighs or only slightly beyond. When properly relaxed the arms will rise to the horizontal in back or sometimes even slightly above.

2. Demonstrate, presenting a side view to the class.

Progression

This will be noticeable mainly in increased ability to release tension in the extensor muscles and in the elimination of jerky movements. It may be necessary to work individually outside of class with a few pupils who have difficulty in learning to "let go." They are the ones who need relaxation the most, and a little extra help is very important.

EXERCISE 4. TRUNK TWISTER.

Purpose

To increase flexibility of the shoulder girdle, spine, and thorax.

Execution

1. Position. Stand with feet at least 18 inches apart, toes straight ahead, right arm bent and placed behind the back so that the back of the hand rests lightly against the lower back, the left arm hanging at the side (Fig. 5, A).

2. Movement. Swing the left arm forward to shoulder level with the palm down (Fig. 5, B), then immediately fling it horizontally sideways and backward as far as possible, twisting the trunk and head to the left as far as possible at the same time (Fig. 5, C). Return in reverse order. After four repetitions change to the other side by swinging the left arm behind the back at the end of the downward movement and then without a pause or change in the rhythm start the right arm forward. Alternate sides every four twists for three or four times.

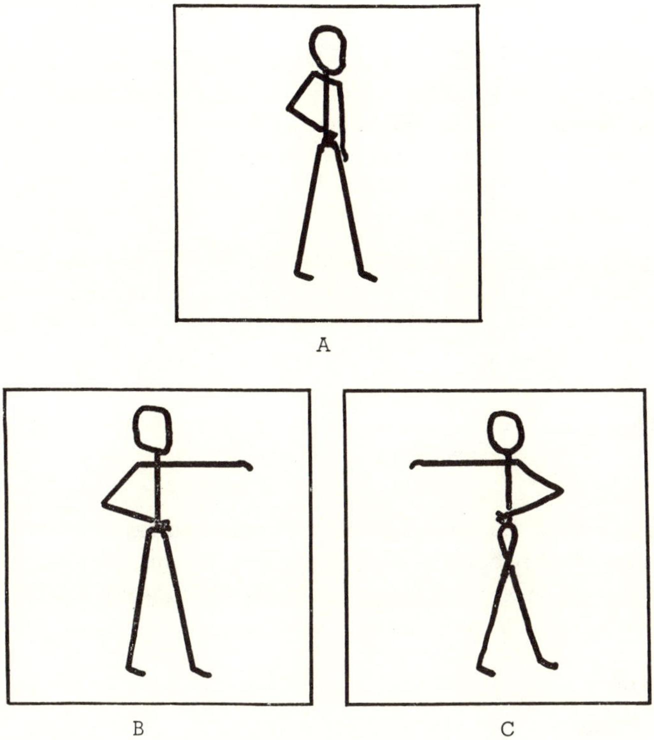

Fig. 5. Trunk Twister.

3. Setting the Rhythm. Establish the rhythm by counts and descriptive terms, such as: "forward-back-forward-down, two-back-and-down, three-fling-and-relax, four-fling-and-change," then repeat the count for the other side, "one-back-and-down, two-fling-and-relax" and so on, varying the use of terms which are descriptive of either the direction or force of the movements. Once the rhythm is established let the class carry on alone.

Cautions

1. The eyes should follow the hand on the backward fling to obtain the maximum twist of the head and trunk.

2. The feet, including the heels, should remain flat on the floor and the knees should be kept straight throughout.

3. The arm should remain at shoulder level on the backward movement and not be allowed to drop downward.

4. The movements should be quite slow and deliberate to obtain the maximum range. Fast movements cause a tension in the shoulder muscles which may result in slight injury or pain. **Never speed up the rhythm.**

5. Demonstrate, presenting a front view to the class.

Progression

The important progression is in the increased mobility of the spine, thorax, and shoulder girdle and the ability to immobilize the pelvis and legs and thus localize the movement in the upper trunk and arms. For variation rather than progression, alternation of sides may be made after each six or eight twists.

EXERCISE 5. TRUNK LIMBERING

Purpose

To increase the flexibility of the spine and legs. (Stretch the muscles at the back of the upper leg, commonly called the hamstrings, and the lower back muscles and the lateral muscles of the trunk. Review Figure 2.)

Execution

1. Position. Stand with feet at least 18 inches apart, toes straight ahead.

2. Movement. This exercise includes movements done in three directions in a series. Movement in each direction is described separately. (a) Stretch the arms out sideways at shoulder level and bend the trunk downward as far as is possible while keeping the knees completely extended. Hold the head up so it is in line with the spine and keep the upper back straight. From this position gradually force the trunk farther and farther downward by a series of small rhythmical forcible stretching movements. Do not let the knees bend. (Fig. 6, A.) (b) Lift the trunk to the upright position, place the right arm in a curved position overhead with the left arm bent and the palm placed over the left hip joint. Bend the trunk to the left side as far as possible, pushing the pelvis and hips toward the right, and letting the head and right arm hang downward to the left. The whole body forms a curve like the letter C. From this position try to force the side bending farther and

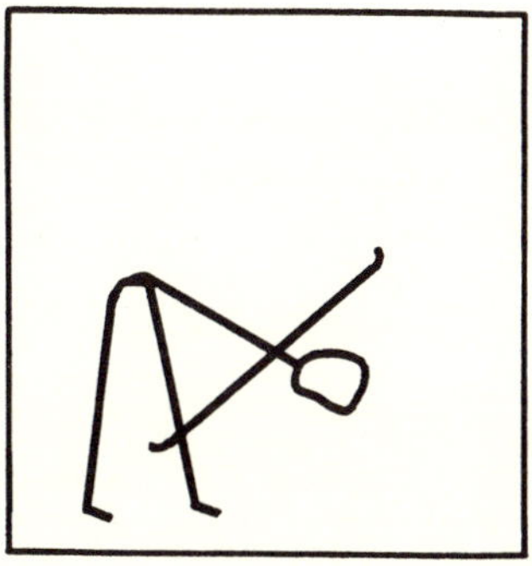

A

B C

Fig. 6. Trunk Limbering.

farther to the left by a series of small rhythmical forcible stretching movements. Do not let the knees bend. (Fig. 6, B.) (c) Bring the trunk to the upright position again, reverse the position of the hands and arms and repeat the forcible side bending to the right side. (Fig. 6, C.)

The movements continue in a series, first downward, then left, then right, then downward again, et cetera. There are several ways of repeating the series, such as

a. 8 downward, 8 to the left, 8 to the right and so on for several series.

b. 4 downward, 4 to the left, 4 to the right and so on.

c. 8 downward, 8 to the left, 8 to the right, then 4 downward, 4 to the left, 4 to the right, then 2 in each direction and finish with one in each direction.

3. Setting the Rhythm. Counting the number of forcible stretching movements in each direction is probably the best manner of setting the rhythm in this exercise since the rhythm is moderately rapid.

Cautions

1. The knees should be kept rigidly straight throughout the downward movement or there is no stretching of either the hamstring or lower back muscles. Beginners tend to bend the knees. Remind the class at frequent intervals to keep the knees straight. Tell them they will feel a pull either behind the knees or in the lower back or perhaps in both places.
2. The movements should not be forced too much, especially with beginners. They should be done as easy bobbing springs with very little lifting of the trunk.
3. During the side bending the arms should be kept in the suggested position and utilized as weight levers to make the bending more vigorous.
4. The side bending should be done straight to the side with the trunk squarely forward throughout. There should be no turning of the head, shoulders, or pelvis in the direction of the bend.
5. Pushing the hips and pelvis sideward away from the direction of the bend aids in keeping the balance and makes the movement more forceful and efficient.
6. The breath should not be held. Breathing should be normal.
7. Demonstrate, presenting a front view to the class.

Progression

In this exercise progression will be obtained only as the movements are properly executed with the knees kept straight and with no rotation of the trunk. The progression will be in increased mobility of the spine as observed by the increased range of movement in each of the three directions. As the individual or group learns to release the tension of those trunk muscles which it is desirable to stretch, there will also be noticeable a more definite relaxation. It is only when the performer knows that there is an uncomfortable "pull" associated with most flexibility exercises and has learned to accept this feeling as desirable, that progress will be observed. Start with a few repetitions only, to avoid undue muscle soreness, and gradually increase the number, but probably a maximum of sixteen to twenty movements in each of the three directions is sufficient to acquire and maintain an adequate degree of trunk flexibility.

EXERCISE 6. FULL SQUATS OR SQUAT JUMPS

Purpose

To strengthen the foot, leg, and thigh muscles. (The extensor muscles of the hip, knee, and ankle joints.)

Execution of Full Squats

1. Position. Stand erect, with one foot slightly ahead of the other so the heel of the forward foot is about opposite or a very

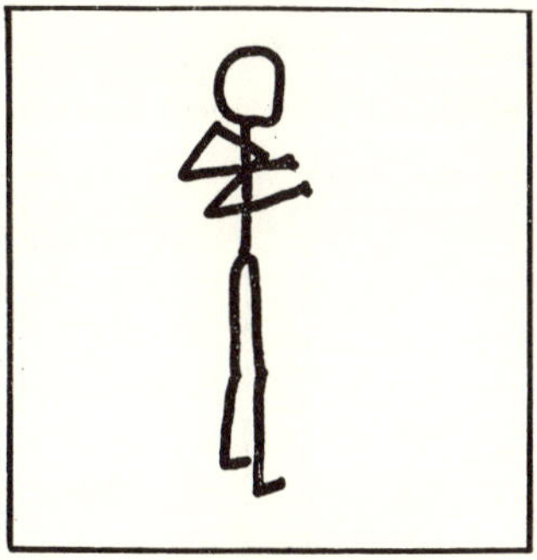

A
Correct
Starting position.

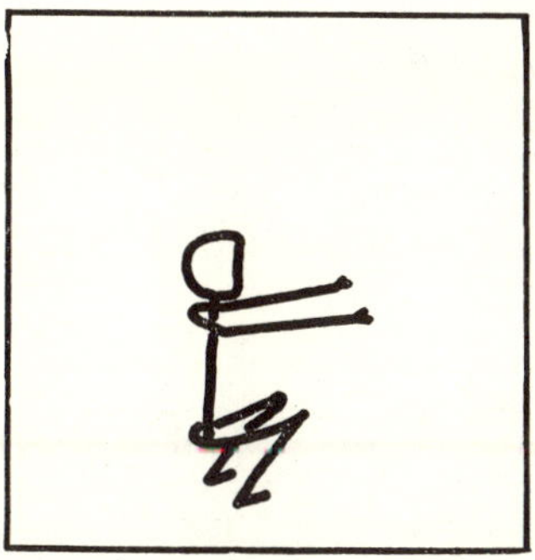

B
Correct
Trunk erect.

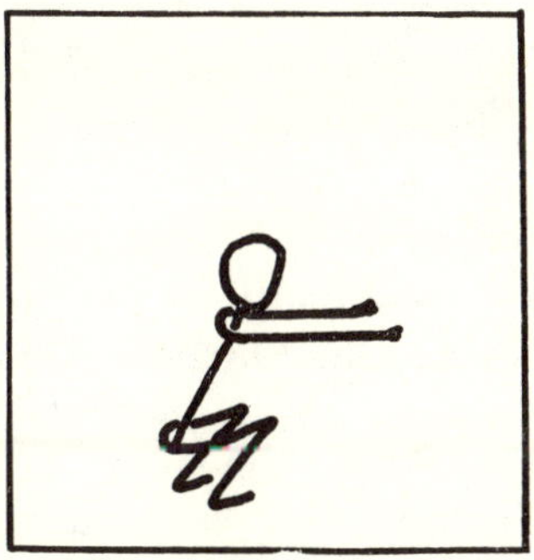

C
Wrong
Trunk bent forward.

Fig. 7. Full Squats.

little ahead of the toes of the back foot. Arms are bent with elbows close to the sides, hands in front of the shoulder joints or slightly lower (called arms in thrust position). (Fig. 7, A.)

2. Movement. Squat down until the buttocks rest on the heel of the back foot and simultaneously extend both arms forward at shoulder height. (Fig. 7, B.) Return to erect position immediately with the arms bent again. The arm movement is used merely as a means of aiding in keeping the balance. Allow pupils to reverse foot position at will. Repeat ten to seventy-five times.

Execution of Squat Jumps

1. Position. Stand erect with the left foot about 8 inches in front of the right foot, the hands placed on top of the head.

2. Movement. Squat down until the buttocks touch the right heel and then immediately bounce and spring upward until the knees are straight and the feet just off the floor (Fig. 8).

Fig. 8. Squat Jumps.

Drop down again to a full squat but with the position of the feet reversed, the right foot about 8 inches forward. Continue bouncing and springing upward, reversing the position of the feet on each jump. Repeat six to sixteen times.

3. Setting the Rhythm. In both variations of this exercise the count indicating the number of repetitions should be given as the upward movement is in progress because it is this action which strengthens the extensor muscles of the hip, knee, and ankle joints. The words "up," "spring," or just numbers may be used.

Cautions

1. It is inadvisable to do Squat Jumps during the menstrual period. Full Squats, however, may be done at any time. Therefore it is suggested that those individuals who are menstruating do Full Squats when the remainder of the class are doing Squat Jumps.

2. Emphasize balance and light landings in Squat Jumps with stress upon good ankle, knee, and hip action.

3. The knee is rather easily injured. Therefore, it is advisable to warn any pupils who have had a knee injury to be very cautious about doing Full Squats. It is probably unwise to allow them to do Squat Jumps at all. The same applies to pupils who complain about their knees hurting. Have these students omit this exercise entirely or do only a few, depending upon the history of the injury. Remember, however, that bending the knees with the weight supported upon two knees results in much less strain than when the weight is supported upon one bent knee at a time as in lunges, charges, or alternate knee bending.

4. Demonstrate, presenting a front view to the class.

Progression

There are three reasons for including both Full Squats and Squat Jumps, two of them having to do with progression. First, it takes more strength to spring into the air than merely to extend the knees without a jump. Therefore Squat Jumps are progressively more difficult than Full Squats. Secondly, Squat Jumps make a greater demand upon the heart-lung mechanism, and therefore should logically come later. Third, by using both exercises as suggested above under cautions, the girls learn that once having progressed to where they can be done, Squat Jumps are preferable except during the menstrual period or at times of stress or convalescing from illness and that instead of omitting this exercise at that time they should do the less strenuous Full Squats.

EXERCISE 7. SIT-UPS

Purpose

To strengthen the abdominal muscles.

Execution

1. Position. Lie on the back on the floor with the knees bent to a right angle, feet on the floor and separated about 18 to 24 inches, feet supported. The position of the arms is varied for progression such as, (a) stretched forward toward the knees, (b) folded on the chest, or (c) with hands clasped behind the head.

2. Movement. Raise the trunk forward-upward off the floor with a definitely curling movement in which the head is raised off the floor first, then the shoulders, and last of all the lower back region of the spine. (Fig. 9, Arm Position 1.) The return movement should be one of uncurling in which the head and shoulders are kept as far forward as possible while the lower back is pushed down on the floor first, then the shoulders are lowered, and last of all the head. In doing the exercise from position (b) with the arms folded on the chest or position (c) with hands clasped behind the head, each trunk-raising movement ends with a twist, touching the right elbow to the left knee on the first Sit-Up, then the left elbow to the right knee on the next Sit-Up (Fig. 9, Arm positions 2 and 3). Repeat ten to fifty times.

3. Setting the Rhythm. Since it seems very difficult to keep a group together in this exercise, it is suggested that there be no attempt to keep them in unison. Give a starting signal and then urge each individual to do as many as she can at her own rate of speed. Have her count the number she does and try to increase the number gradually from week to week.

Cautions

1. The use of trunk-raising movements as a means of strengthening the abdominal musculature has been criticised, and justly so. The difficulty, however, lies not so much in the movement itself as in the way it is performed. It may be done so that it overdevelops the hip flexor muscles, especially the psoas, and strains rather than strengthens the abdominal muscles, or it may be performed so that strong abdominal action is obtained. Some teachers urge that students do trunk raising with the chest leading and the spine kept as nearly

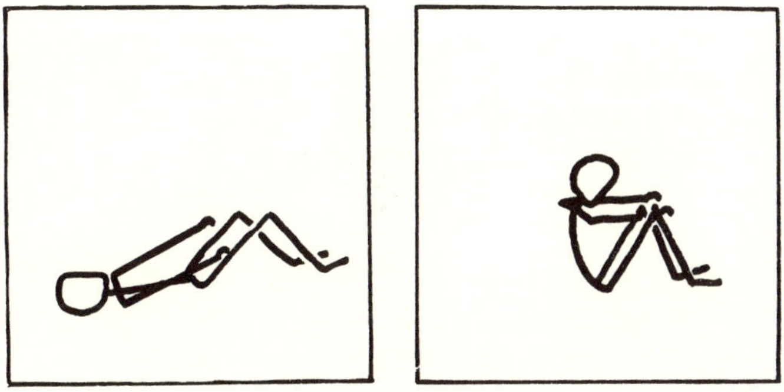

Arm position 1

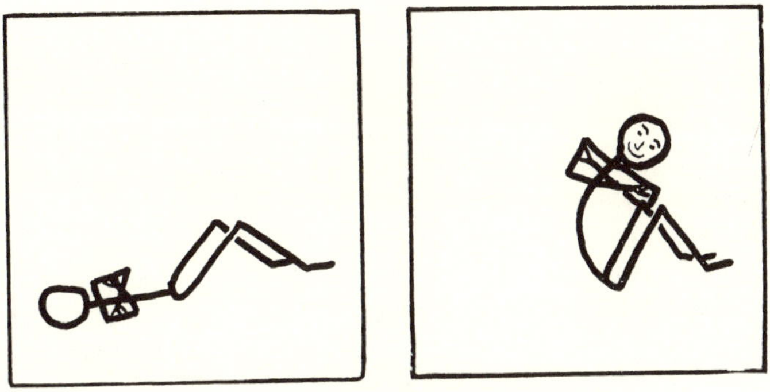

Arm position 2

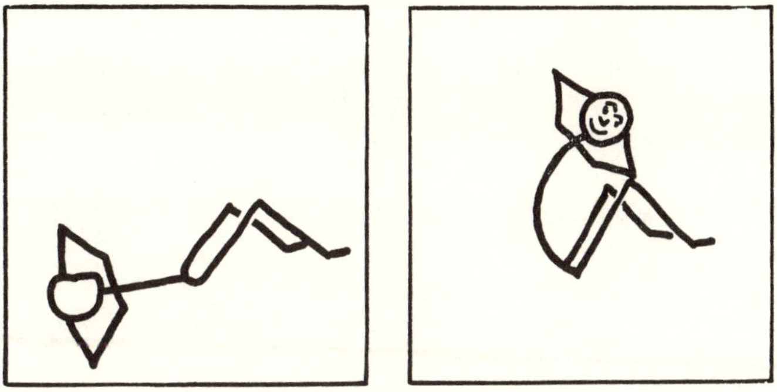

Arm position 3

Fig. 9. Sit-Ups.

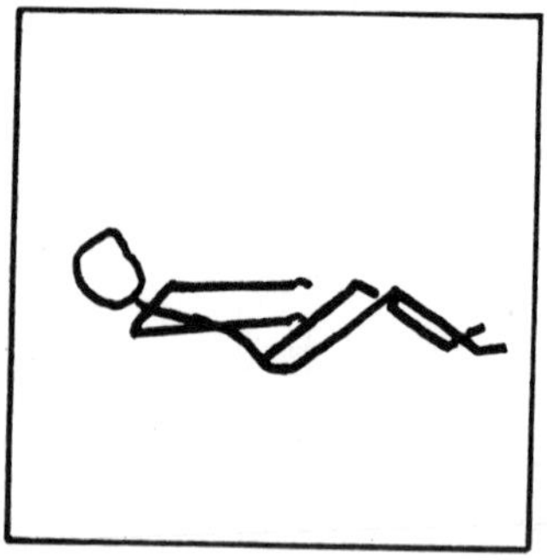

Fig. 10. Wrong performance of Sit-Ups. Legs too straight and psoas muscle doing most of the work as indicated by hollow lower back.

straight as possible (Fig 10). Such a performance probably takes a minimum of action for the abdominal muscles and a maximum of work for the psoas muscle (a muscle which brings the thigh and the abdomen closer together). The abdominal muscles are normally trunk flexor muscles and to bring them into maximum action the upward movement must involve a bending or curling forward of the spine. The return movement should be one of lengthening contraction of the abdominal muscles in which an attempt is made to extend the spine by an uncurling movement. It helps students to get the correct action if they have the idea of lowering each vertebra to the floor one at a time starting with the lowest lumbar vertebra and ending with the head.

2. The authors would like to advocate that the feet should not be anchored under an object or held down by a partner. If the feet are held down, the psoas muscle aids in raising the trunk or sometimes (as in many ballet or acrobatic dancers who have very strong hip flexors) does the work alone. There are a few girls, however, who, because of a disproportionately long trunk and short legs or heavy trunk and thin thighs and legs, find it impossible to do this movement when the feet are not supported. A compromise is suggested—that the feet be held down **lightly,** giving just enough support to offset the two disadvantages mentioned above but not enough to make it possible for the psoas muscle to do the work alone. Have the girls work in pairs, lying on the floor feet to feet. When the correct position is assumed, the knees bent to a **right angle** and

feet separated 18 to 24 inches, the partners must be close enough that each girl may place her right foot on top of the left foot of her partner. This will give each girl a light support on one foot. The Sit-Ups are then done alternately, one girl doing the upward movement as her partner is going down and vice versa. This not only provides a lighter support but saves more time than if each girl does the work alone while her partner holds her feet.

The next caution gives the criterion for determining whether the major part of the work is done by the psoas or the abdominal muscles.

3. The instructor can tell by a side view of the performance whether it is the psoas or the abdominals that are doing most of the work. The psoas pulls forward on the lower back region of the spine and the individual who raises the trunk with this part of the spine in a position of hyperextension (or sway back) is using the psoas muscle to perform most of the work. (Fig. 10.) If the lower back is flexed and the spine curved forward (Fig. 9) then the abdominal muscles are doing the work.

4. This exercise need not be omitted at the menstrual period but it may be wise to curtail the number usually done. Students with menstrual difficulties should be urged to do only one-third to one-half the usual number the first two days of menstruation and very probably the last two days just preceding the onset of the flow.

5. Hernia, chronic appendicitis, or any abdominal tenderness or soreness (other than muscle soreness) are indications that abdominal exercise is not advisable.

6. Demonstrate, presenting a side view for the first position with arms stretched forward because the curling of the trunk is the most important factor to be demonstrated. Later when done with arms folded on the chest or with hands clasped behind the head, demonstrate, presenting a front view to show the twist of the trunk alternating to the left and right.

Progression

Each individual should start doing the Sit-Ups in the easiest position, which is with the arms straight and stretched forward toward the knees, so that there may be concentration upon the correct muscle action to insure the use of the abdominal muscles instead of the psoas muscle. When fifteen or twenty Sit-Ups can be correctly done with the arms in this position, progress to the next step which is with the arms folded on the chest and the movement done with a twist. This twist insures action of the external and internal oblique abdominal muscles which, as well as the rectus abdominus, need to be strengthened. When fifteen or so can be done in this way, change the arms to a position of hands clasped behind the head. Continue with them in this position, progression from then on being one of increasing the number of repetitions and the speed of execution.

EXERCISE 8. HIP RAISING

Purpose

To strengthen the extensor muscles of the lower back and hip joint.

Execution

1. Position. Lie on the back with the arms on the floor by the sides, the knees and hips flexed and the feet flat on the floor.

2. Movement. Lift the hips (the whole pelvic region really) up into the air. Lower to the floor again (Fig. 11). Repeat ten to twenty times.

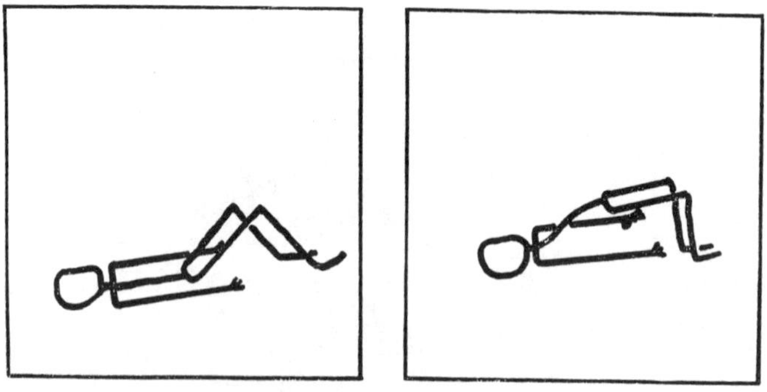

Fig. 11. Hip raising.

3. Setting the Rhythm. Establish the rhythm by such suggestive terms as; "up-down", or "one-and, two-and, three-and".

Cautions

1. Be sure pupils lift the pelvis up far enough to insure good muscle action of the lower back and hip extensors.

2. There are a few individuals who will get the action too far up the back with the lower back lifted in a lordotic (sway back) position. These students should be taught how to roll the pelvis up at the front by a contraction of the abdominal muscles and down at the back by pinching the buttocks together and should roll the pelvis into this position just preceding each hip-raising movement. Women need lower back strength for many homemaking duties, therefore the lower back muscles should not be neglected. Many exercises which would strengthen these muscles have the disadvantage of putting the lumbar spine in a position of excessive hyperextension, but the position in which this exercise is executed with the knees bent excludes the possibility of an extreme lordotic position of the spine.

3. Demonstrate, presenting a side view to the class.

Progression

Do not have the class try for the maximum number of repetitions in this exercise. The lower back needs to be strong enough to prevent low back aches or low back strains but care must be taken that the strength of the abdominal musculature always equals or exceeds the strength of the lower back muscles to prevent lordosis (sway back). Therefore, a maximum of twenty back-arching movements is suggested.[4]

EXERCISE 9. HIP ROLL

Purpose

To strengthen the abdominal muscles, especially the oblique abdominals. (Review Fig. 1.)

---

[4]Many teachers of physical education believe that there should be no stress upon strengthening the lower back muscles on the theory that strong muscles in this region will increase the all too prevalent tendency toward lordosis. There should not be too much stress, however, on sparing the back muscles at the expense of strengthening them. Weak back muscles result too frequently in sacro-iliac strains and other back difficulties. The remedy is not to neglect the back muscles but to train for correct habits of pelvic tilt and strengthen both the abdominal and lower back muscles.

Execution

1. Position. Lie on the back on the floor, arms out sideways at shoulder level with palms pressing against the floor. Bend the knees to the chest, feet free of the floor, knees kept together (Fig. 12, A).

A

Elementary exercise with knees bent.

B

More strenuous exercise with legs straight.

Fig. 12. Hip Roll.

2. Movement. Swing the knees to the left to touch the floor on the left side, then swing them across the body close to the chest to touch the floor on the right side, and continue swinging from one side to the other. Repeat ten to thirty times.

3. Setting the Rhythm. Verbal suggestions for this exercise are as follows; "left-and-right, and two-and-right, and three-and-right, and four-and-over," and so on.

Cautions

1. The feet should be kept free of the floor throughout and the knees close to the chest.

2. The knees should be kept together.

3. The palms of both hands should be kept in contact with the floor throughout even if a stretching pull is felt in the shoulders.

4. Pupils should breathe normally. Urge them not to hold the breath.

5. Demonstrate, presenting both front and side view to the class.

Progression

Increase the number of repetitions until the maximum is reached and the muscles are sufficiently strong to do the exercise with the legs extended straight upward at right angles to the trunk, feet and legs together and knees straight (Fig. 12, B). Start with about six swings in this position and increase gradually to twenty.

EXERCISE 10. PUSH-UPS

Purpose

To strengthen the arm and shoulder girdle muscles.

Execution

1. Position. Take a position face down toward the floor in which the body is supported in a slanting position upon the hands and knees. The feet are up in the air with the knees bent at right angles. The arms are extended and straight with the hands directly under the shoulder joints, fingers parallel to length of body. The body must be in a straight line from head to knee (Fig. 13, A).

2. Movement. Lower the body slowly until the chin touches[5] the floor and then push up again to the starting position without

[5]Normally if an individual keeps the head in line with the spine so the body is in a straight line from head to knee the chest would touch the floor before the chin. In order to keep the abdomen or thighs from touching the floor, however, the body is not in an absolutely straight line—there is a slight angle at the hip joint. This means that the hips are very slightly higher than the shoulders and this inclination of the trunk means that it is possible to touch the chin to the floor and still keep the head very nearly in line with the spine. The authors have found that touching the chest to the floor too frequently results in either an exaggerated angle at the hip joint or in allowing the abdomen to touch.

Fig. 13. Push-Ups. A and D show correct performance. B and C show wrong starting positions; B with arms bent, body bent, and hips raised, and C with back sagging and hollow and thighs on the floor. E and F show wrong lowering of the body; E with hands too far forward and abdomen and thighs touching the floor, and F with hands too far back and thighs touching the floor.

relaxing. Abdomen, hips, or legs must not touch the floor and there must be no motion anywhere in the body other than in the arms and shoulders (Fig. 13, D). Repeat six to fifty times.

3. Setting the Rhythm. Each up and down movement in this exercise should be done at first upon a separate signal with sufficient pause in each position to allow for correction. When done in rhythm such terms as "down—and up, down—and two, down—and three," may be used. The rhythm should be quite slow at first until the movements are accurate, then the speed of execution should gradually be increased.

Cautions

1. There must be no sagging of the back. Warn pupils that the abdomen must be tensed and kept firm to prevent this (Fig. 13, C).
2. Tell pupils there is a very slight angle at the hip joint to prevent the abdomen and thighs touching the floor. This angle is so slight, however, (5 degrees at maximum) that the trunk and thighs are very nearly in a straight line (Fig. 13, D).
3. The head must be held up and not allowed to sag downward.
4. The head must not be thrust downward in order to touch the chin to the floor, but kept up and the whole body lowered until just the chin touches.
5. Watch for the trunk being lowered by bending at the hip joint so that the hips are left up in the air as the shoulders are lowered. The whole body must be lowered as one unit, the abdomen and thighs coming very close to the floor but not touching (Fig. 13, B, D).
6. There must be no undulating motion of the trunk as the body is pushed up (Fig. 13, C).
7. Tell pupils to breathe normally. They should not hold the breath.
8. Demonstrate, presenting both front and side view to the class.

Progression

It is exceedingly important that the position and movement for this exercise be learned accurately right at the start. It is suggested therefore that each individual lower the body each time only as far as she can control the body position. Thus the progression would be (1) by increasing the depth of the dip from perhaps 3 or 4 inches at first until finally the chin touches the floor (2) by increasing the number of Push-Ups whether partial or full and (3) by increasing the speed of execution.

Experience has taught that starting on the floor with a push-up and allowing the whole body to touch the floor on each movement is not as wise a procedure in progression as the one suggested above. If the abdomen is allowed to touch the floor even momentarily it is relaxed too much and the upward movement too frequently becomes an undulating one of lifting the shoulders first by the arms and then lifting the abdominal region by contracting the abdominal muscles.

By using partial Push-Ups as suggested here each individual progresses at her own rate, some of the class doing partial Push-Ups while the remainder do full Push-Ups.

EXERCISE 11. WING LIFTS[6]

Purpose

To strengthen the posterior muscles of the neck and upper back. (Review Fig. 2.)

Execution

1. Position. Lie face down on the floor with the head tilted backward, the elbows bent and hands behind the neck, elbows lifted and kept off the floor, toes touching the floor.

2. Movement. Lift the head and upper back until the chin is about 12 inches from the floor or until the breasts just barely leave the floor, no more than one inch above the floor, however. Return to starting position (Fig. 14). Do not lower the elbows and do not allow the toes to come up off the floor at any time. Repeat twenty to one hundred times or as many as possible in one minute.

3. Setting the Rhythm. Accuracy of movement and muscles used is more important than speed in this exercise. Therefore it is desirable at first to keep the class together and the rhythm

[6] See The Official Physical Training Program of the Women's Army Corps: You Must Be Fit. Washington, Infantry Journal, 1943, p. 57.

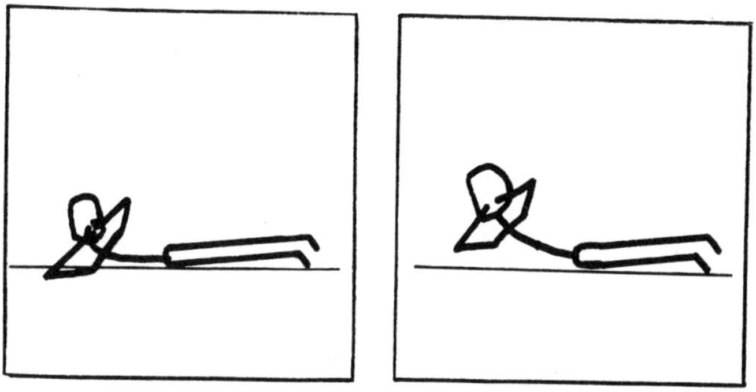

Correct
Uses neck and upper back muscles only.

Wrong
Lift is too high, uses lower back muscles.

Fig. 14. Wing Lifts.

slow until the movements are correctly learned. When the coordination is well established, emphasis should be placed upon speed of movement and when this time comes there should be no attempt to keep the group together. It is suggested that the maximum number of lifts which can be done in one minute is one good way of motivating this exercise. In this case the following signals could be used: "Wing Lifts [pause] ready [pause] go! [pause] ten seconds [pause] twenty seconds [pause] thirty seconds [pause] forty seconds [pause] fifty seconds [pause] stop!"

Cautions

1. The elbows must not touch the floor at any time.

2. Action should not consist merely of tilting the head backward and forward by a bend in the neck which tips the chin up and down. The head should be tilted backward as far as possible for the starting position. The movement then becomes one in which the back or crown of the head describes an arc backward-upward as though attempting to touch the heels. Another method which helps some individuals to get the correct muscle action is for the instructor to stand in front of the class in such a position that all pupils can see her. On each lift of the upper back the pupils are told to lift the head high enough to see the face of the instructor. Both of these methods have given success, but best results have been obtained by the

admonition, "Try to make the back of your head touch your heels."

3. Definite motion should be observed in the dorsal spine involving some degree of extension.

4. If the upper back is lifted too high the motion is the result of a contraction of the lower back muscles which is not desirable. The purpose here is to strengthen the upper back muscles and train them to hold the dorsal spine in a position of extension. Therefore the movement must be localized in the upper back and the back lifted only so far as to cause the breasts to just barely leave the floor.

5. Watch for those who merely lift and lower the elbows in a sort of flapping motion which lifts the upper trunk off the floor by action of the latissimus dorsi muscle but does not result in any motion of extension in the dorsal spine. The elbows should be lifted as high as possible and then kept up throughout.

6. The toes must remain in contact with the floor at all times. If the feet are lifted it means the lower back and hip extensor muscles are being used. Insisting on the toes remaining down will help in localizing the movement in the upper trunk and eliminating the lower back muscles.

7. When rest is needed before the one minute, or other time limit used, is up, allow them to rest with the chin on the floor but with the elbows kept off the floor.

8. Demonstrate, presenting a side view to the class.

Progression

The progress at first will be in learning the correct muscle action. This is very important. The individuals who have formed the habit of slouching and allowing the upper back muscles to be relaxed will be the ones who have greatest difficulty in learning to use these muscles. They are also the ones who for this reason need it the most. Give them all the individual help and attention possible. Keep the rhythm slow and the group together until the coordination is well learned. Then progression should be by increasing both the speed and number of lifts. It is here that it is advisable to begin setting a time limit upon the exercise and to have each individual attempt to do as many lifts as possible within this time. Progression in this exercise then is, first, learning the correct coordination, and second, increasing the number of repetitions and, third, increasing the speed.

EXERCISE 12. ABDOMINAL PUMPING

Purpose

To increase circulation of the blood throughout the pelvic region, particularly to relieve or prevent dysmenorrhea.

Execution

1. **Position.** Lie on the back with the knees and hips flexed and the feet flat on the floor. One hand rests lightly on the lower abdomen.

2. **Movement.** Slowly push upward on the abdominal wall until it is distended as far as possible, then slowly pull in on the abdominal wall until it is retracted as much as possible. The movement should be slow, smooth, and continuous with no jerks or pauses and should be a muscular movement entirely with no relationship to the rhythm of breathing. Breathe normally and naturally. Having the hand rest lightly upon the abdomen seems to aid in localizing the movement and indicates the amount of movement. Repeat ten times.

3. **Setting the Rhythm.** No attempt should be made to keep the group in unison. The rhythm should be slow and smooth. The teacher can tell when it is time to give the signal to stand for the next exercise by watching and counting the number done by some one girl.

Cautions

1. This exercise should never be done with the legs extended. Watch to see that the hips and knees are bent so that the abdominal muscles are lax enough to allow for proper movement.

2. A slow, smooth, rhythmically continuous action is necessary—no jerky movements.

3. Beginners may have some difficulty in divorcing the rhythm of the pumping from the rhythm of the breathing, but persistent concentration will soon accomplish it.

4. Demonstrate, presenting a side view to the class.

Progression

There is no further progression in this exercise once it is properly learned so that it is accurately performed. Ten repetitions at one time is usually considered sufficient to accomplish its purpose.

EXERCISE 13. FOOT ROCKER

Purpose

To strengthen the muscles supporting the longitudinal arch.

Execution

1. Position. Stand erect, feet about 3 inches apart, and very slightly pigeon-toed.

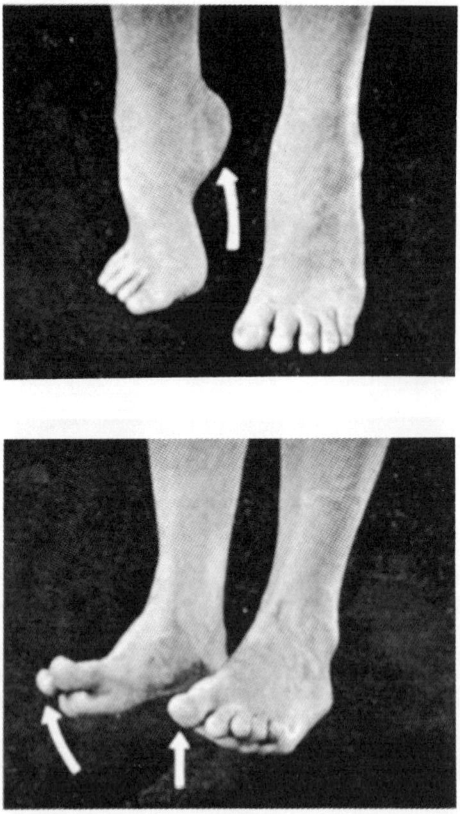

Fig. 15. Foot Rocker. (WAC Manual, WD Training Circular No. 40, 16 June 1944, Field Manual 35—20. Courtesy of Department of the Army.)

2. Movement. Rise up on the toes, lifting the heels from the floor, then lower the heels and lift the front part of the foot from the floor (Fig. 15). Repeat twenty to fifty times.

3. Setting the Rhythm. Establish the rhythm by the use of numerals which also indicate the number of repetitions.

Cautions

1. As the weight is transferred from the heel to the ball of the foot, it should be carried through the midline of the foot and not along the inner border to prevent pronation.
2. As the ball of the foot is lifted the great toe should be lifted slightly higher than the little toe side of the foot so the soles of the feet are turned very slightly toward each other.
3. An erect position of the body should be maintained throughout. Warn to avoid swaying the trunk or hips forward and backward.
4. For the individual with flat feet or extremely low and weak arches, this exercise is not recommended. It is recommended as a preventive procedure and for correcting cases of slight to moderate pronation accompanying slightly low or weak longitudinal arches. Individuals with very weak or low arches need special and individualized attention.
5. Demonstrate, presenting both front and side view to the class.

Progression

Progression is obtained through gradually increasing the number of times the movements are executed until the maximum is reached and then maintained there.

TEACHING SUGGESTIONS SPECIFIC FOR CONDITIONING EXERCISES

Suggestions for teaching exercises in general are given in Chapter IX, but while we are discussing specific exercises of a conditioning series it seems wise not to separate from their description the teaching suggestions which pertain to them specifically as a series. Therefore the following presentation is given in this chapter.

Warming-Up

Since the conditioning exercises of this chapter are designed to be used in an everyday, keep-fit program, they should follow the physiological principle of warming up gradually to the peak of effort and then tapering off. The first five exercises of the thirteen are of this warm-up type; the next six take nearly the maximal effort, some more than others, and the last two take less exertion.

Order of Exercises

Other than for warming-up there is no special required order of exercising the various parts of the body. The important thing is to use them in rotation so that when one exercise demands maximal or near maximal effort from one muscle group the next exercise provides for minimal exertion of this group by shifting the major work to other muscles, thus allowing the first group time to recover from fatigue. This principle has been adhered to in the series of conditioning exercises given in this chapter. The untrained and inexperienced teacher would do well to retain this order.

Also, to save unnecessary shifting from a standing to a lying position or from standing with feet apart to feet together, the exercises have been grouped when possible so that starting positions are similar. The first exercises are done in the standing position, then there is a change to exercises done on the floor, and finally a return to the standing position.

Changing From One Exercise To The Next

It is recommended that the change from one exercise to the next be done in the following manner:

The teacher stations herself at the front of the room with the class facing her while the exercises Arm Circling through Full Squats (or Squat Jumps) are being executed. Upon the signal "Partners facing, on back—down!" or "Partners facing—lie down!" the class changes to a recumbent position with sides toward the teacher. In other words, two lines of girls will face each other, then lie down feet to feet. The correct way of doing a "Sit Fall" (which is described later) may be used here. After Sit-Ups, Hip Raising, and Hip Roll are done, a signal such as

"With heads toward front of room," "To a position face down—over!" or simply "With heads front—roll over!" will get them ready for Push-Ups.

After Push-Ups and Wing Lifts are completed the class again rolls over onto their backs for Abdominal Pumping. The next signal is either "Facing front, in erect position—stand!" "Foot Rocker—begin!" or "Facing front—Stand up!—Foot Rocker—go!"

The exercises Arm Circling through Full Squats lend themselves well to a rhythmically continuous motion of one exercise flowing into the next without cessation of movement. If the proper signals are correctly timed greater satisfaction results for the class than if movement is stopped after each exercise while the next one is named. It works out as follows. While the class is doing the movements of Arm Circling the words "Body Swing" are spoken as the right arm is circling for the second time. The class continues circling the right arm and the signal "Now!" or "Go!" or "Begin!" is given as the seventh circle is started. The arm completes this circle and on the eighth count swings forward-upward to an overhead position, the left arm swings upward also from its position on the hip, the left foot remains in its forward stride position, and the movements of Body Swing start and continue even when the teacher says "Placing left foot sideward, Trunk Twister." After a pause of about one second, as the arms and trunk start on an upward extension, the signal "Go!" or "Begin!" is given and the class swings smoothly into the Trunk Twister exercise. The words "Trunk Limbering" are said as the arm swings forward for the beginning of the last twist to be done that day (whatever number has been decided upon) and the signal "Now!" or "Begin!" comes as the arm starts downward toward the side at the end of this last twist. The signal "Full Squats . . . begin!" is similarly timed so movement flows smoothly from Trunk Limbering to Full Squats.

First Presentation Of An Exercise

When first presenting an exercise the following procedure should be meticulously observed:

1. Arrange the class in couples so that the students may assist the teacher in criticising first performances. This may

be done easily by having the first, third, fifth columns of students do the movements with criticism from the partners in second, fourth, or sixth columns.

2. Name the exercise, then describe the movements while demonstrating them, follow by repetition of the name which is to be used exclusively once the class becomes familiar with the particular exercise the name is supposed to call to mind. For example, when first presenting the exercise to be used in this series by the name of "Trunk Twister" it is necessary to explain what is meant by the term for there are many different exercises which are trunk twisters and the pupils need to be told this and then informed about the particular trunk twister that is to be taught to them. Once this exercise is explained to them and they have memorized the movements involved, the teacher and pupils can thereafter speak of it merely as "Trunk Twister."

3. State its purpose briefly as given in the outline at the beginning of the chapter.

4. Demonstrate it accurately a second time so as to leave a clear, vivid image in each student's mind.

5. Call attention to common faults to avoid but stress what to do rather than what not to do.

6. Announce which group is to try the exercise first, warning the other group to be ready to observe and criticise the performance of their partners.

7. Restate the name of the exercise using only its brief form such as "Sit-Ups." Follow the name immediately by the appropriate signal for the first group to start action. Appropriate signals are those such as "Go!," "Begin!," or "One!" if movements are to be counted and each one done only on signal.

8. Have the pupils do the movement only a few times as this first trial is for accuracy and not for physiological stimulation.

9. The assisting group observes and criticizes the performers individually while the teacher criticizes the class in general.

10. Now that their first performance has been criticized, have the first group do the exercise once more with both partner and teacher criticism.

11. Repeat Numbers 7 to 10 having the second group try the exercise with the first assisting and criticizing.

This procedure should be followed whenever a new exercise

to be used frequently thereafter is first presented to the class. On subsequent days the partner idea may be repeated as desired, especially with the more difficult exercises, but this should not be continued too long for it cuts the exercise period of each pupil in half.

Accuracy In Exercising

For the sake of acquiring accuracy of movement and an understanding of the purposes of an exercise it is advisable to continue for quite a number of lessons to state the purpose of the exercise with either a brief review demonstration or a very short suggestive description which will recall it to the student's mind. This should be continued until the teacher is sure that all (a) know the exact movements that accompany the name of the exercise, (b) understand the purpose of the exercise and (c) can do it accurately. These things must not be sacrificed for the sake of "working up a sweat." Organic stimulation will come readily enough once knowledge and accuracy of movement are satisfied. This is the groundwork that must not be neglected if later results of classwork are to be good. It is the painstaking scale practicing of the musical novice.

Techniques of Demonstrating

Some exercises should be so demonstrated as to present a front view to the class, some a side view, and a few perhaps should be demonstrated twice, showing both front and side view. When demonstrating an exercise to be done on the floor, be sure the class is sitting up. Do not expect them to be able to see from a recumbent position. Always demonstrate where all can see and if it is a movement to be done to the left by the class, demonstrate it to the right when facing the class.

If the demonstration has been sufficiently clear and accurate, all of the exercises may safely be started immediately in slow rhythm with the exception of Push-Ups, as previously explained. Each up and down movement in this exercise should be done at first upon a signal with sufficient pause in each position to allow for correction.

Unison

During the first few weeks unison is not as important as accuracy of performance. As the movements become familiar and more rhythmical, unison should be stressed, except as noted for certain exercises.

Elimination Of Pauses Between Exercises

The pauses between exercises which are necessary when first presenting and demonstrating them should be eliminated as rapidly as possible until the exercises, finally memorized as a series, can be performed from start to finish with little or no cessation of movement. It will take a few weeks to bring a class to this stage but it is worth working towards for once accomplished the conditioning series can be run off quickly, much organic stimulation is obtained by the pupils, and, moreover, time is saved for other class projects.

Suggestions in General

Some brief but definite suggestions are appended to aid the teacher in making the use of these exercises not only more purposeful and successful but also more educational.

1. Memorize the names of the exercises in the correct order and have the class learn to recognize them by name.

2. As time goes on the class also should know the names of the exercises in the correct order. Give them a short written quiz of about five minutes two or three times during the semester to check on the accuracy of their knowledge and the efficiency of your teaching.

3. Be sure that you know the purpose of each exercise in the series as stated briefly in the listing on page 65. In addition, be able to interpret the results of each exercise in terms of correction or prevention of faults such as prominent abdomen, forward head, round shoulders, and pronated feet. Impart both purpose and interpretation to the class frequently enough to insure their knowledge and understanding. An occasional quiz will help impress them with the importance of such insight.

4. Decide before the class meets on the number of repetitions of each exercise which are to be done and keep a written account to insure proper progression. Table I gives a suggested progression scale.

Table I
PROGRESSIONS FOR EXERCISES IN CONDITIONING SERIES*

Week of the Semester	1	2	3	4	5	6	7	8	9	10	11	12	13	14	15
Warmer-Upper	1½m	1½m	1½m	1½m	1¾m	1¾m	2m	2m	2¼m	2½m	2¾m	2¾m	3m	3m	3m
Arm Circling	8	8	8	16	16	16	16	16	16	16	16	16	16	16	16
Body Swing	6	8	8	10	12	14	16	16	16	16	16	16	16	16	16
Trunk Twister	8	8	12	12	12	12	16	16	16	16	16	16	16	16	16
Trunk Limbering	8	12	12	15	15	15	15	15	15	15	15	15	15	15	15
Full Squats	10	15	20	25	30	35	40	45	50	Squ 10	at Ju 15	mps 20	25	30	35
Sit-Ups	8	10	12	15	18	20	22	25	27	30	32	35	40	45	50
Hip Raising	10	12	14	16	18	20	20	20	20	20	20	20	20	20	20
Hip Roll	10	12	15	20	25	with 6	legs s 8	traight 10	12	14	16	18	20	20	20
Push-Ups	8	10	12	15	18	20	22	25	27	30	32	35	40	45	50
Wing Lifts	10	15	20	25	35		maxim	um in	one	minut	e				
Abdominal Pumping	10	10	10	10	10	10	10	10	10	10	10	10	10	10	10
Foot Rocker	20	20	25	30	35	40	45	50	50	50	50	50	50	50	50

*The numbers after the name of each exercise represent the number of repetitions of that exercise which should be done for the week of the semester recorded at the top of each column.

5. In determining the number of times to repeat an exercise, decide on what the majority can do and warn the weaker students to stop an exercise when they should. Urge these students to increase their capacity by individual work outside of class.

6. Do not have beginners use too much force in the flexibility exercises. The result is apt to be an antagonistic protective tension in the muscles it is desirable to stretch. It is important to learn first to release all tension and resistance in these muscles, and only when this has been accomplished should force be used.

7. When this series is used in classwork in school it may be found both valuable and interesting to teach the group how to do a correct "Sit Fall"[7] as the method of shifting from a standing position to lying on the back (as going from Full Squats to Sit-Ups). Also, a correct and easy method of rising from the back to a standing position again can be taught (after Abdominal Pumping to stand for Foot Rocker).

8. Do not forget that these exercises are useful only as they contribute to conditioning the body and to developing body control and efficiency in movement. Therefore, be constantly aware of the postural pattern and teach not only toward the ability to control the various body masses in movement and keep them in balance, but toward kinesthetic awareness and the knowledge of how it can be accomplished. This means an understanding on the part of the pupil of the principles underlying correct (efficient) body mechanics and how they apply to the movements of the exercises. For a listing of these principles see Chapter VII.

9. Encourage the students to use these exercises at home and to teach them to other members of the family.

The Use Of Music

Most of the exercises of this conditioning series can be done to the accompaniment of a musical instrument, records, or even singing. Whenever possible it is probably advisable to do so for part of the time because it adds interest and stimulation

---

[7]U. S. Office of Education: Physical Fitness Through Physical Education for the Victory Corps, p. 62. U. S. Office of Education, Victory Corps Series, Pamphlet Number 2.

and makes the work more satisfying. It aids some individuals in acquiring better relaxation. It saves the monotony of too continued use of the voice in setting and keeping the rhythm.

Decide beforehand on the number of measures of music to be used for the correct number of repetitions of each exercise, so that you may direct your attention in class toward individual help and correction. Use numbers of repetitions which will correspond to musical phrases. It is displeasing to stop the music in the middle of a phrase.

The tempo of the music for each exercise must be correct to obtain the maximum benefit. When first using music, have the accompanist play two or three measures as a cue to establish the rhythm in the minds of the students. Then do the first few repetitions of the exercise with the class to be sure the rhythm is really correct. Later, when both accompanist and teacher are sure of the correct tempo neither of these cautions will be necessary, and the music can continue without pause from one piece to the next as the class changes exercises.

Whenever possible use the music of familiar songs, and after the movements are well learned and do not require too much concentrated attention, encourage the class to sing while exercising. In cases where a piano and pianist, or records and music box are not available, music may be provided entirely by singing. This latter is advisable only after the exercises are quite familiar and performance is accurate.

Musical Selections to Use

There are endless possibilities as to the particular musical selections to use for these conditioning exercises. Any teacher with a smattering of musical training and a good sense of rhythm and timing and ability to adjust the music to the type of body movement involved will be able to make her own selection, but for those who desire guidance or who wish to save time the authors suggest the use of five books which have been tried out and have proved of value in this work. They are as follows:

Book 1. Dietrich-O'Donnell: Notes for Modern Dance. New York, A. S. Barnes & Co., 1937.

Book 2. Frost, Helen: The Clog Dance Book. New York, A. S. Barnes & Co., 1924.

Book 3. Frost, Helen: Clog and Character Dances. New York, A. S. Barnes & Co., 1924.

Book 4. Ryan, Grace: Dances of Our Pioneers. New York, A. S. Barnes & Co., 1939.

Book 5. Weir, Albert E.: Masterpieces of Piano Music. New York, Mumil Publishing Co., 1918.

These books are used for the specific exercises of the conditioning series as follows:

1. **Warmer-Upper** (May be done with or without music.)
   Book 1, pages 9 and 10 are suggested.
   Tempo M. M. ♪ = 134 to 144.

2. **Arm Circling**
   a. Book 3, page 21. Country Dance
      Tempo M. M. ♪ = 84.
      The arm makes four circles to each measure of the music, or, if the music is played twice as fast, the arm makes two circles to each measure of the music.
   b. Book 3, page 36. Jack Frost
      Tempo M. M. ♩ = 84.
      The arm circles twice to each measure of music.

3. **Body Swing**
   Book 1, page 16. Rocking
   Tempo M. M. ♩ = 42.
   One complete movement (down and up) to each measure of music.

4. **Trunk Twister**
   a. Book 4, page 164. The Campbells Are Coming
      Tempo M. M. ♩. = 58.
      Two measures of music are needed for one complete movement forward, back, forward, and down.
   b. Book 3, page 19. Sleigh Bells
      Tempo M. M. 𝅗𝅥 = 58.
      Two measures of music are needed for one complete movement.

5. **Trunk Limbering**
   a. Book 2, page 25. On the Railroad
      Tempo M. M. ♩ = 72.
      One bob or bend to each beat, four bobs or bends to each measure of music.
   b. Book 1, page 11. Fall, Lift (bottom of page)
      Tempo M. M. ♩ = 72.
      One bob or bend to each beat, four bobs or bends to each measure of music.

6. **Full Squats**
   a. Book 3, page 25. Rig-a-Jig
      Tempo M. M. ♩. = 58 to begin with and gradually progress to 72 or faster.
      One complete movement (down and up) to each measure of music.
   b. Book 2, page 27. On Deck
      Tempo M. M. ♩. = 58 to begin with and gradually progress to 72 or faster.
      One complete movement (down and up) to each measure of music.

7. **Sit-Ups**
   Book 4, page 56. Captain Jinks
   Tempo M. M. ♩. = 80.
   Two measures of music are needed for one complete movement, one measure for the trunk raising and a second to lower the body again.

8. Hip Raising

a. Book 1, page 29. Leg Whip
Tempo M. M. ♩ = 54.
One complete movement (up and down) to each measure of music.

b. Book 2, page 10. Dixie (first 16 measures only)
Tempo M. M. ♩ = 54.
One complete movement (up and down) to each measure of music.

c. Book 1, page 30. Leg Swing Advancing
Tempo M. M. 𝅗𝅥. = 54.
Two measures of music are needed for one complete movement (up and down).

9. Hip Roll

a. Book 1, page 29. Leg Swing
Tempo M. M. ♩ = 69 when done with knees bent.
Tempo M. M. ♩ = 44 when legs are kept straight.
It takes two measures of music for one complete movement (to right, up, to left, up).

b. Book 3, page 23. Swanee
Tempo M. M. ♩ = 69 when done with knees bent.
Tempo M. M. ♩ = 44 when legs are kept straight.
One complete movement (to right, up, to left, up) to each measure of music.

10. Push-Ups

Book 1, page 4.
Tempo M. M. ♩. = 44.
One complete movement (down and up) to each measure of music.

11. Wing Lifts

a. Book 2, page 17. Yankee Doodle
Tempo M. M. ♩ = 40 to begin with and gradually progress to 108.
Two lifts of the upper back to each measure of music.

b. Book 3, page 25. Rig-a-Jig
Tempo M. M. ♩. = 80 to begin with and gradually increase the tempo.
One complete movement to each measure of music. This is good music for the beginning weeks when accuracy is most important and the rhythm is slow. When the tempo is rapid the Yankee Doodle music is preferable.

12. Foot Rocker

a. Book 1, page 12. Pedal and Run
Tempo M. M. ♪ = 92 and gradually increase to 100.
One complete movement to each measure of music (heels up, down, toes up, down).

b. Book 5, page 380. La Cinquantaine (The Golden Wedding) (first 40 measures without repeats).
Tempo M. M. ♩ = 92 and work up to 100.
One complete movement (heels up, down, toes up, down) to each measure of music.

c. Book 2, page 32. The Georgian Male Quartette (first 20 measures only and then repeat)
Tempo M. M. ♩ = 92 and work up to 100.
One complete movement (heels up, down, toes up, down) to each measure of music.

d. Book 3, page 15. For Practice of 3's and 3-3-7
Tempo M. M. ♪ = 92 and work up to 100.
One complete movement to each measure of music (heels up, down, toes up, down).

TIME REQUIRED FOR THE SERIES

If the conditioning exercise series suggested in this chapter is used as a whole the amount of time required for its execution will be considerably more the first few lessons, when demonstrations and careful corrections are necessary, than later on when the movements are familiar. It may require all of the first two class periods to teach the series adequately. After this the time requirement gradually lessens until 15 to 20 minutes is sufficient, the variation dependent upon the ability of the class in terms of numbers of repetitions of each exercise. The time remains fairly constant after the first five or six lessons, the augmented knowledge and understanding of the movements and the increasing smoothness of execution balancing the time required for doing the exercises a greater number of times. It has been found that twenty minutes is sufficient for taking roll call, executing a three-minute run or jumping rope, and doing the full series of exercises.

MODIFYING THE SERIES

In a class in body mechanics and conditioning twenty minutes is not too much time to spend in the development of adequate muscle strength, muscle balance, endurance, and flexibility. As a preliminary conditioning series for a sports class, however, or a swimming class, this may be too much time, particularly if the class period is only thirty-five minutes in length. A group of students who are familiar with the conditioning series given here can do a slightly modified series as given below in nine minutes, exclusive of time for roll call and preliminary run.

1. Warm-up run .... 2 minutes
2. Opening order for exercises .... 1 minute
3. Exercise Series .... 9 minutes
   a. Arm Circling—16 times
   b. Body Swing—8 times
   c. Trunk Limbering—16 in each direction
   d. Full Squats—50 times
   e. Sit-Ups—up to 25 times
   f. Push Ups—up to 24 times
   g. Wing Lifts—one minute
   h. Foot Rocker—50 times

Total amount of time .... 12 minutes

It will be noted that the above series omits one trunk flexibility exercise (Trunk Twister) but retains the other one (Trunk Limbering). Hip Raising is omitted. Hip Roll is omitted but Sit-Ups are kept for strengthening the abdominal muscles. Abdominal Pumping is not included upon the assumption that the complete series has been used for one semester and the resulting familiarity with this exercise, including knowledge and understanding of its purpose and use, should fulfill its educational aim.

CHAPTER V

SUPPLEMENTARY EXERCISES FOR SPECIFIC NEEDS

The physical activities suggested in this chapter are not intended as substitutes for any of the conditioning exercises of Chapter IV but as supplementary exercises to be used in addition to them in special cases. No one individual would find it necessary to use all of the exercises suggested here.

Each pupil should be given an understanding of her physical status, her posture pattern, her strengths and weaknesses. This is obtained from the material of the medical examination, the physical examination, the body mechanics test, the strength tests, and any other source of information available. From all of these sources the teacher obtains knowledge of a pupil's strengths and weaknesses: For example, a pupil may have fair strength of trunk muscles and passably good posture and endurance but have weak and pronated feet. For this individual the daily use of the conditioning series should be supplemented by an additional exercise or two chosen from the section on foot exercises. Another student may be about average in all respects except for poor alignment of the lower back and abdominal region resulting in lordosis (sway back), relaxed abdominal muscles, and constipation. She needs special exercises chosen from the section on abdomen and lower back and the section on constipation.

In addition to material for a pupil's own immediate use this chapter should supply information about the correction of faulty postural patterns or functional disturbances which she may pass on to her family or friends or may find useful herself at some future time. The authors consider it wiser to spend the time which can be allotted to this purpose in gaining a thorough knowledge of a few good exercises, rather than a scanty understanding of many. Therefore only a few exercises are included under each heading.

OBJECTIVES

Each of the exercises which follow has been included because it fulfills one or more of the following objectives:

1. To stretch short tissues which pull some segment of the body out of good alignment and balance.
2. To strengthen weak muscles, especially those which need to be strengthened to obtain good muscle balance of opposing muscle groups (equalizing exercises).
3. To provide kinesthetic awareness of good mechanical balance and substitute correct habit patterns for faulty postural patterns.
4. To provide additional exercises for the sake of variety where supplementary work is needed.

FORWARD SHOULDERS AND FLAT CHEST

The usual results of a prolonged habit of relaxing the muscles of the upper back and allowing the shoulders to droop forward and the chest to be depressed are weakened and stretched upper back muscles, especially the upper erector spinae and rhomboids, and foreshortened chest muscles, particularly the pectoralis major and minor. (See Figs. 1 and 2.) To overcome this condition the short muscles should be stretched and the weak muscles strengthened.[1] The exercises which follow fulfill these purposes.

Exercise 1

On hands and knees, fling one arm sideways upward as far as possible, twisting the head and trunk at the same time so the eyes follow the hand. Keep the supporting arm straight

[1]A good motivating device for suspension exercises which are excellent for this condition is a stout rod placed across a doorway near the top leaving merely comfortable grasp clearance. The device has been used at Oberlin College where Dr. Gertrude Moulton had such rod attachments placed in several doorways in the women's gymnasium and in the women's dormitory. The rods, being removable, were not in the way and the rod holders were carefully placed so as not to interfere with doors. Girls were urged to make daily use of the rods by swinging through doorways.

Fig. 16. Exercise 1 for forward shoulders and flat chest.

(Fig. 16). As the arm swings downward relax it and let it bend at the elbow and swing across the chest. Repeat several times on one side before changing hands. Alternating four on each side is good.

Exercise 2

Sit on the floor, knees bent upward, feet flat on the floor in front of the hips (hook sitting), hands on floor, palms down, at the side of the body. Fling one arm forward-upward as far as possible keeping the elbow straight, arm passing close to the ear and going as far past the ear as possible, palm facing forward (Fig. 17). Replace the hand and repeat with the other arm. Then fling both arms at the same time. Continue flinging one arm, the other arm, then both arms, using enough force to feel a real stretching sensation in the region of the shoulder joint.

Exercise 3

Kneel, sit back on the heels, trunk bent forward and downward until the forehead rests upon the floor, arms stretched backward, hands resting palms upward on the floor, thumbs toward the body. Lift the head and upper trunk until the spine is straight, keeping the lower abdomen and thighs close together. Rotate the arms outward at the same time so thumbs move downward and then outward to a position of thumbs pointing upward, palms facing out (Fig. 18). Relax and return to starting position. For greater difficulty and variation the

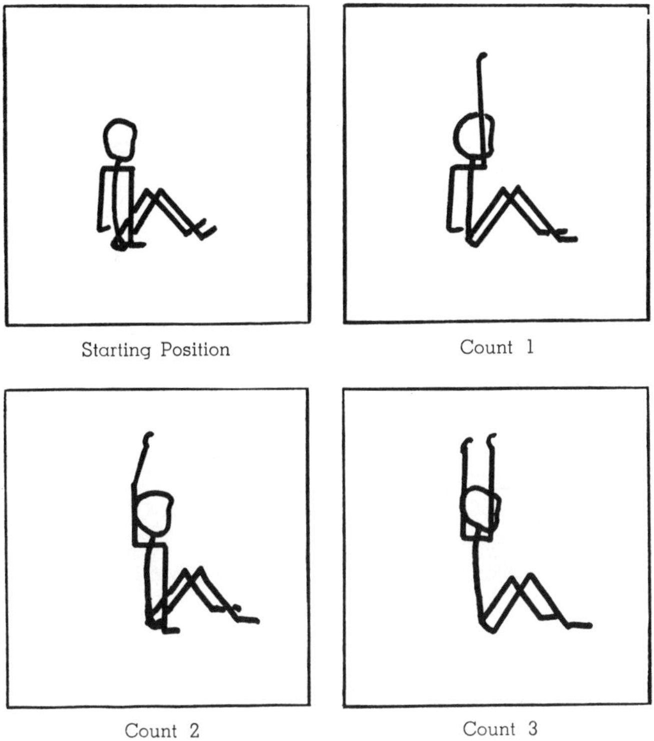

Fig. 17. Exercise 2 for forward shoulders and flat chest.

hands may be placed back of the neck and the elbows lifted upward as far as possible as the trunk is lifted, or the arms may be stretched sideward at shoulder height, palms down, and the whole arm lifted.

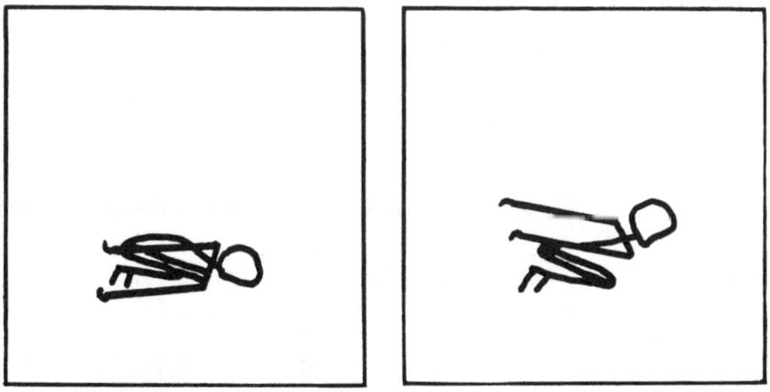

Fig. 18. Exercise 3 for forward shoulders and flat chest.

Exercise 4

Sit on the floor with the soles of the feet touching in front of the body, knees spread sideways as far as possible, hands firmly grasping the ankles. Bob the trunk forward-downward four times, forcing it by pulling on the ankles (Fig. 19, A). Then

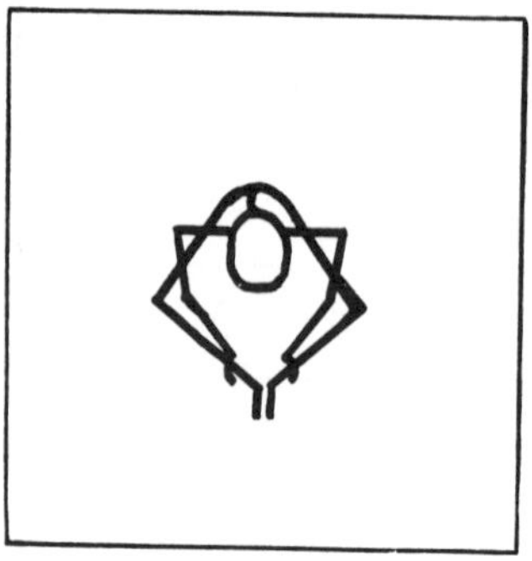

A. Top view forward-downward movement.

B. Sideview of back extension.

Fig. 19. Exercise 4 for forward shoulders and flat chest.

stretch the head and trunk upwards, attempting to arch the whole back, hands still grasping the ankles. The resistive pull of the hands on the ankles should aid in lifting the chest (Fig. 19, B). Hold for four counts. Alternate four bobs downward with four counts of back extension. The bobbing downward aids also in stretching the lower back muscles and if the knees are forcibly spread sideways, hip flexibility results.

ABDOMINAL AND LOWER BACK REGIONS

The construction of the lower spine and pelvis is such that any increase or decrease in the tilt of the pelvis involves a corresponding increase or decrease in the curve of the lower back (Fig. 20). The angle at which the pelvis is carried is due to habit, to the relative strength of the abdominal and lower back muscles, or to the relative length of the lower back, hamstrings and psoas muscles, and the abdominal muscles. Most people who do not carry the pelvis at the correct angle tilt it too far forward and this results in an exaggerated curve of the lower spine (sway back), lengthened and weakened abdominal muscles, and foreshortened psoas, lower back, and hamstring muscles. This means insufficient support for the visceral organs which frequently results in painful menstruation, constipation, or other functional disorders. The exercises which aim at correction of lower back and abdominal abnormalities, therefore,

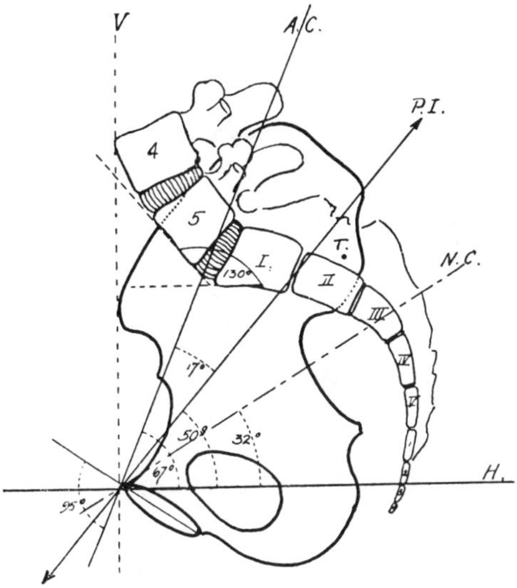

Fig. 20. Pelvic tilt. A pelvic angle of 50 to 60 degrees is considered normal. If the upper part of the pelvis is tilted farther forward, the curve of the lumbar spine is forced farther forward also, resulting in hollow back and protruding hips. (After Fick from Morris: Human Anatomy. P. Blakiston Sons & Co., Inc., Philadelphia.)

should do one or more of the following: (1) strengthen the abdominal muscles, (2) stretch and elongate the psoas muscle, (3) stretch and elongate the lower back and hamstring muscles or (4) substitute correct habit patterns of pelvic tilt for faulty ones.

Exercise 1. Psoas Stretching

(This muscle does not show in Figure 1 as it is buried too deep in the pelvis. It extends from the front sides of the lumbar vertebrae downward behind the pelvic organs and is inserted at the inside of the upper end of the femur [thigh bone]. When it contracts it brings the trunk and thigh at an angle with each other.)

Assume the position shown in Figure 21 with the weight on the forward foot. Holding this position bob the hips up and down, keeping the back leg as straight as possible. When this

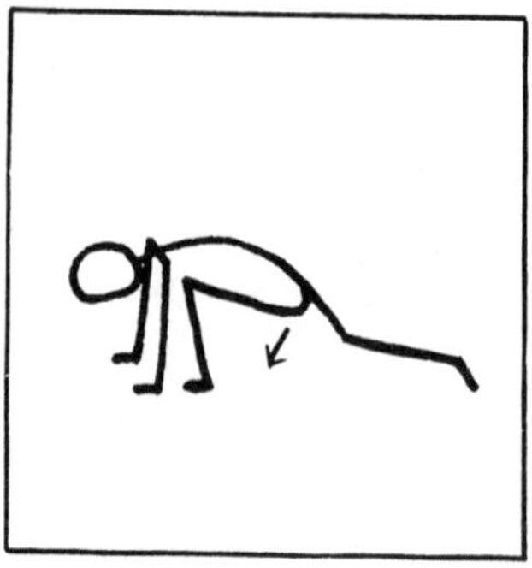

Fig. 21. Psoas Stretching.

exercise is executed correctly there is felt a strong pull on the front of the thigh at hip joint level.

Exercise 2. Lower Back And Hamstring Stretcher

Stand with the feet apart. Bend the trunk downward toward the outside of the right foot then bob or bounce the trunk up and down moving it around in a circle by touching the floor

Fig. 22. Lower back and hamstring stretcher.

first outside the right foot, then forward, then forward but farther to the left, and finally outside the left foot, then raise trunk upward to erect position, swing across and bend downward again on the right side to repeat (Fig. 22). Keep the knees rigidly straight throughout the exercise. After several repetitions reverse directions.

Exercise 3. Abdominal Curl

Lie on the back on the floor, knees bent upward, feet flat on the floor (hook lying), arms folded on the chest. Lift the head and shoulders off the floor so the upper trunk curls forward.

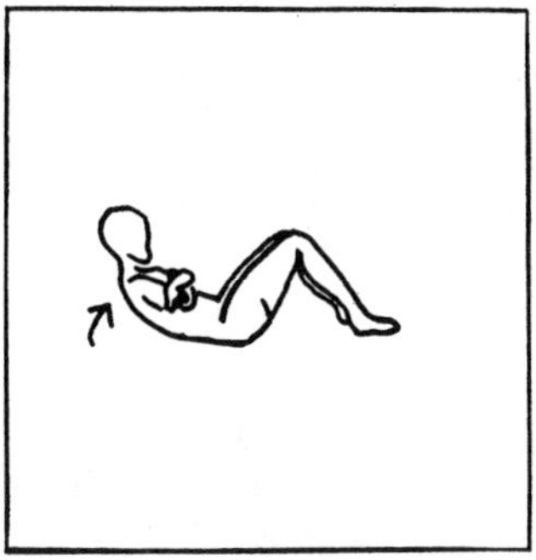

Fig. 23. Abdominal Curl. (After Paul C. Williams, M.D.)

Keep the lower back on the floor (Fig. 23). Return to starting position. This is easier than the sit-up exercise of the conditioning series and may be used where a mild abdominal exercise is indicated. (See Fig. 23.)

Exercise 4. The "V" Sit-Up

Lie on the back on the floor with the arms out sideways at shoulder level. Raise the trunk and extended legs upward

Fig. 24. The "V" Sit-Up.

simultaneously until they form a V and hold for several seconds or as long as possible (Fig. 24). This is a very strenuous abdominal exercise.

Exercise 5

To strengthen the oblique abdominals. (Review Fig. 1.) Lie on the floor on the back, arms out sideways at shoulder level, palms pressed against the floor. Raise the right leg straight up in the air, then swing it across the body and up to touch the left hand. Return it to place. Raise the left leg up and swing

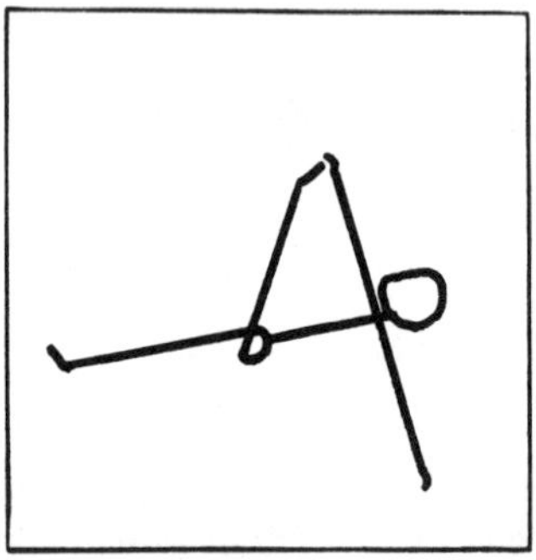

Fig. 25. Exercise for strengthening the oblique abdominals.

it over to touch the right hand (Fig. 25). Repeat, alternating sides. This is a less strenuous exercise than Hip Roll of the conditioning series when done with the legs straight, but is probably about the same as Hip Roll done with the knees bent.

FOR ESTABLISHING CORRECT HABITS OF PELVIC TILT

Exercise 1

Lie on the back on the floor, knees bent upwards, feet on the floor (hook lying). Contract the abdominal muscles pulling the abdomen downward toward the floor. At the same time pinch the buttock muscles together rolling the buttocks upward off the floor. (See Fig. 26.)

Exercise 2

Same as exercise 1 above as illustrated in Figure 26 except that it is done with the legs straight.

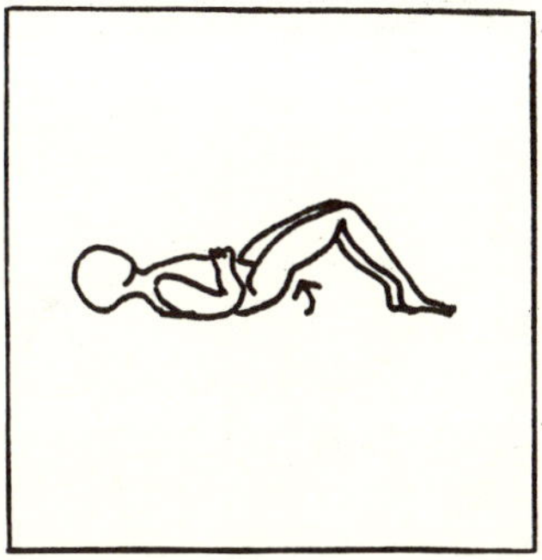

Fig. 26. Exercise for establishing correct habits of pelvic tilt. (After Paul C. Williams, M.D.)

Exercise 3

After the above movement has been mastered while lying on the floor it should then be practiced from a standing position. Each pupil should be taught the correct angle at which the pelvis should be carried and then given opportunity to practice it until she has acquired enough knowledge and sufficient "muscle feel" (kinesthetic awareness) to make self correction possible.

FOR SACRO-ILIAC STRAINS

Sacro-iliac strain is all too common among girls and women. It is due to weak musculature or to ligaments that are too unyielding. To correct or prevent this condition, one needs to strengthen the lower back muscles and also to stretch the ligamentous structures. (Additional information is given in Chapter VIII.)

Exercises for ligaments that are too unyielding are:

Exercise 1

Stand with the feet apart and knees straight. Twist to the left, place the right hand on the outer border of the left thigh and keeping knees straight run the hand as far as possible down the leg—keep stretching farther and farther down—until it hurts. Repeat, running the left hand down the outer border of the right leg (Fig. 27).

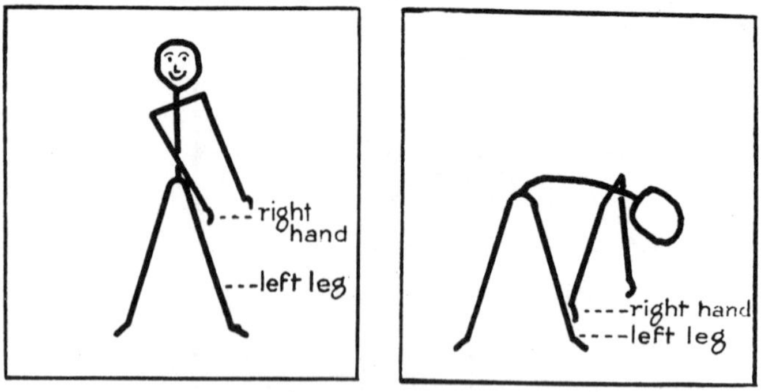

Fig. 27. Exercise 1 for sacro-iliac region.

Exercise 2. Hip Winging

Beatrice Woodcock has the following to say in behalf of this exercise:

> Hip-winging is frequently overlooked, perhaps because sacroiliac movement is considered so small that it is not worth restoring . . . Lost sacroiliac movement means a wooden, springless walk. More important still, in a boy it means a back vulnerable to industrial injuries, while in a girl it means a pelvis vulnerable to the hazards of child-birth. Moreover, stiffness in one part results in secondary stiffness in other parts.[2]

Hip winging is performed while lying on the back on a table with the shoulders held down by grasping the sides of the table or while lying on the floor in such a way as to keep the shoulders flat. The right leg is lifted upward, bent at the knee, and crossed over the left. The heel of the right foot is placed under the left knee from the outer side. The movement consists of pushing the right knee downward toward the floor or table on the left side with the left leg kept straight and the shoulders down. Repeat on the other side.

FEET

The foot is composed of twenty-six bones grouped in such a way that there are three main points which transmit the body weight to the ground: the base of the great toe, the base of the little toe, and the heel. An imaginary line connecting these three points forms a triangle, two sides of which are composed of arches. The metatarsal (transverse or anterior) arch extends

---

[2]Beatrice Woodcock: Scoliosis. Stanford University, California, Stanford University Press, 1946, p. 63.

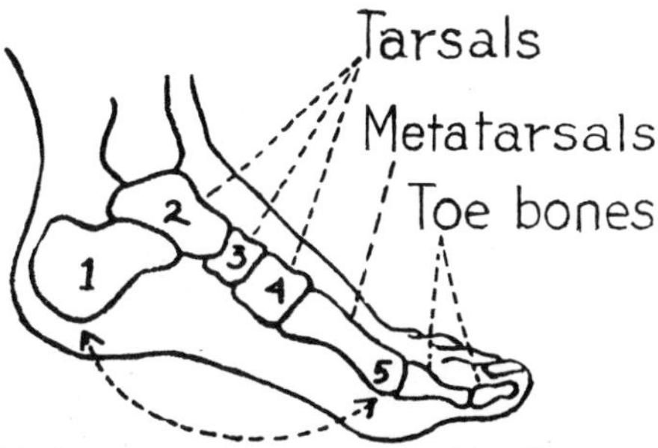

FLEXIBLE SPRING ARCH FROM 1 TO 5.

Fig. 28. The bony structure of the foot as seen from the great toe side. (From Staffs in Consumer and Physical Education, Stephens College: America, We'll Be Walking Now.)

across the ball of the foot from the base of the great toe to the base of the little toe, formed by the anterior ends of the metatarsal bones (Fig. 28). This arch is supported mainly by muscles. The longitudinal arch extends from the base of the great toe to the heel along the inner border of the foot. This arch also is supported mainly by muscles. The third side of the triangle, the outer border of the foot, is supported by ligaments which are next to the strongest in the body.

The weight of the body tends to flatten out the two arches, especially the longitudinal arch. (See Fig. 28.) This is resisted by the muscles supporting the arches but if these muscles are weak then the ligaments holding the bones together take the strain. If this strain continues over a period of time the ligaments gradually become stretched and the normal balance of the foot position is disturbed. The foot tips inward, the inner malleolus (ankle bone) becomes prominent, the tendon of Achilles bends inward just above the heel. This position is called pronated foot, a common foot fault (Fig. 29, B). A low arch in itself, however, does not proclaim a pronated foot. Many people have low arches normally and some even have no arch at all (natural flat foot), yet the foot is strong and otherwise in good condition. The low arch that proclaims the weak foot is the one which is lower than its own normal condition and accompanied by pronation of the foot. If left uncorrected for a period of time it results in a complete flattening of the

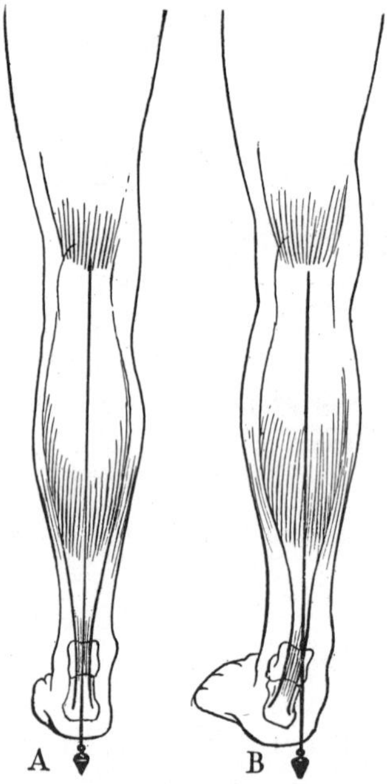

Fig. 29. Plumb line dropped from back of knee. A, normal. B, positive Helbing sign in pronated foot (Lewin: Am. J. Dis. Child.)

longitudinal arch called flat foot, which can be cured only by surgery.

In the intermediate stages of pronation before the ligaments are stretched very much, the judicious use of arch supports supplemented by a daily regimen of good foot exercises will sometimes correct the condition. It is for this individual with weak but not flat feet that the following exercises will be found beneficial.

Exercise 1. Foot Circling

Remove shoes. Sit on stool or chair with one knee crossed over the other or on floor with legs straight, hands under one thigh lifting this leg off the floor. Circle the foot around from the ankle joint, the toes of the right foot moving in a clockwise direction, the left foot in a counter clockwise direction so that in both cases the sole of the foot is facing inward as the toes make the upward swing (Fig. 30). This direction is important.

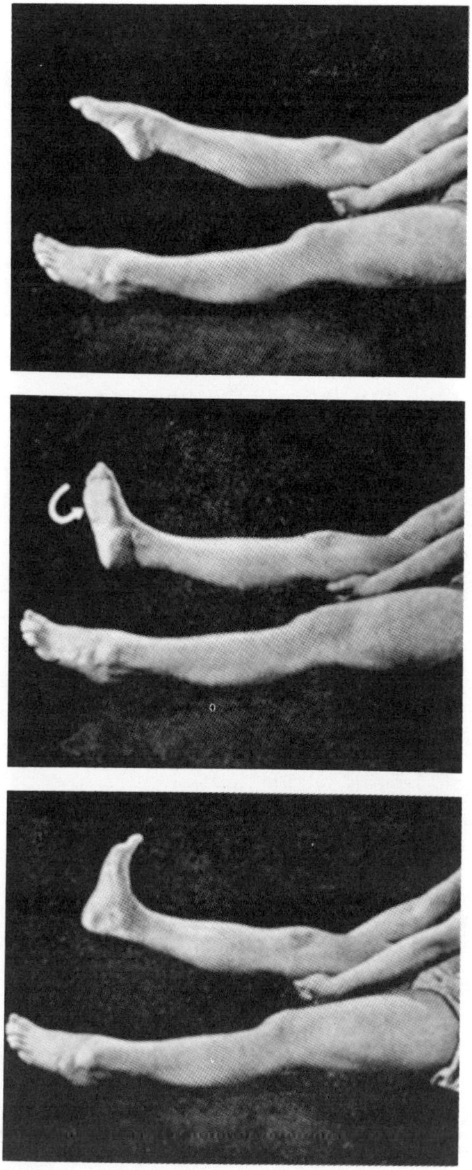

Fig. 30. Foot circling. (WAC Field Manual: Physical Training. Courtesy of Department of the Army.)

Exercise 2

Remove shoes. Sit on the floor with the legs stretched straight out in front, trunk supported by the hands. The feet are separated about 3 inches and held so they are at right angles to

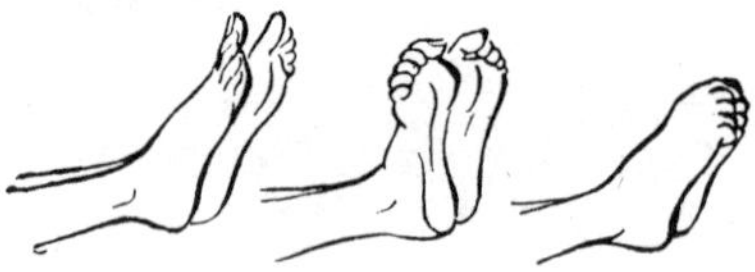

Fig. 31. Exercise 2 for the feet. (From Steinhaus and others: How to Keep Fit and Like It. Permission of George Williams College.)

the floor. In this position curl the toes on count one, keep the toes curled and turn the soles of the feet toward each other on count two (Fig. 31), hold the position on count three, relax the whole foot on count four. Return to starting position and repeat any number of times.

Exercise 3

Remove shoes. Stand so the hands may touch some object for partial support. This may be a wall, window sill, table, or a partner. Two pupils working together may hold each others' hands. Stand with the feet parallel and about 4 inches apart. On count one rise on the heels so the forepart of the foot is free of the floor. On count two curl the toes and, keeping them curled, turn the soles of the feet slightly toward each other, swinging the toes together at the same time. Hold this position on count three. Return to starting position on count four and repeat.

TO TEACH CORRECT WEIGHT BEARING ON THE FEET

Exercise 1

Stand with feet parallel and 2 to 3 inches apart. Press the great toe firmly against the floor and then rotate the thighs outward, keeping the great toe down (Fig. 32). This is accomplished by pinching the buttocks together and contracting the outward rotator muscles of the hip joint. Some pupils understand this better if they are told to turn the legs so that the knee caps turn away from each other. This should lift the longitudinal arch and shift the weight from the inner border of the foot to the center where it belongs.

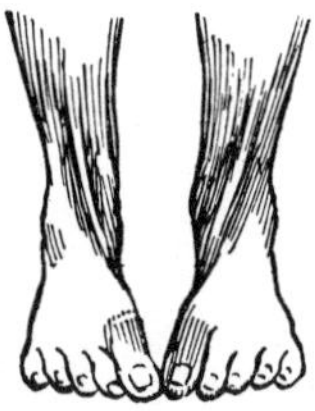

Fig. 32. Exercise for teaching correct weight bearing on the feet. (From Lewin: Posture and Hygiene of the Feet. Courtesy of Funk and Wagnalls Co., Publishers.)

CONSTIPATION

If constipation develops rather suddenly in an individual whose bowel habits have previously been regular, or if it accompanies or follows an infection or disease, a physician should be consulted as to treatment. We are concerned here only with those uncomplicated cases of chronic constipation which are mainly the result of too little exercise, a sluggish peristaltic action, or weak and relaxed abdominal muscles. Cases of this type can often be improved or cured by the right balance of exercise, diet, and rest.

TREATMENT BY EXERCISE

When treatment by exercise is indicated, it should include the following:

1. Participation in active sports or other activities which will stimulate all body functions.

2. Exercises which twist and compress the abdomen and help move the feces by mechanical stimulation.

3. Exercises which strengthen the abdominal muscles.

Several exercises included in the conditioning series of Chapter IV fulfill these objectives. Trunk Twister twists the abdomen, Trunk Limbering compresses the abdomen, and Sit-Ups and Hip Roll strengthen the abdominal muscles. In addition the two following exercises are often helpful because of their compressive action and mechanical stimulation.

Exercise 1. Trunk Circling

Sit on a low stool, hands on hips. Bend the trunk to the right, then circle around to the left pressing the abdomen against the thighs and then raise the trunk upward on the left side to the starting position. Next time start to the left and circle the trunk in the opposite direction. Alternate directions.

Exercise 2

Lie on the back, arms out at shoulder height, legs straight. Twisting the legs to the right, bend the knees and swing them up on the right side of the body to the chest, keeping them as close to the floor as possible without touching it. Keeping the knees bent and pressing the thighs against the abdomen, swing the legs across to the left side almost to the floor (Fig. 33).

Fig. 33. Exercise 2 for correction of constipation. Double knee circling from right to left.

Then, keeping them twisted to the left and just clear of the floor, stretch the legs downward to the starting position. **Never reverse directions** in this exercise because the purpose of it is mechanical pressure on the large intestine in the line of direction in which the feces move, which is upward on the right side (the ascending colon), across the abdomen from right to left (the transverse colon), and downward on the left side (the descending colon).

DYSMENORRHEA

It has been estimated that 45.5 per cent of all women in the United States suffer from dysmenorrhea (painful menstruation) and that 16.6 per cent or about 5.3 million women are actually incapacitated at the time of each menstrual period. This annual loss of work capacity figured at two hours for each menstrual period for each of these five million and more women equals the loss of "an entire year of work by approximately 58,000 women."[3] No wonder that during World War II industry became so aroused over the number of work hours lost by women that it carried on investigations in the hopes of finding a cure and prevention for painful menstruation. The result of several experiments was the salvaging for the war effort of hundreds of thousands of work hours. This should not be the concern of industry, alone, however. The schools, and particularly departments of physical education, should be held responsible for teaching to all junior and senior high school girls the causes, cure, and prevention of dysmenorrhea.

CAUSES OF DYSMENORRHEA

For the purposes of this book the authors deal only with those causes which result in conditions responsive to exercise.

The cause of painful menstruation in many women still baffles the physician. Sometimes it is easily discernible but in most cases it is obscure. The number of young girls and women who have been partially or completely relieved of pain by exercises designed to improve posture, improve the circulation of blood through the pelvic organs, thus relieving congestion, or to stretch shortened fascia of the pelvic region would lead to a tentative diagnosis of poor posture, poor circulation, or contracted fascia as the three major causes.

EXERCISES FOR THE CURE AND PREVENTION OF PAINFUL MENSTRUATION

The most common of the exercises recommended by physicians and used in schools and industry for the prevention and relief of dysmenorrhea are given in the material that follows. Since so high a percentage of girls and women suffer from painful menstruation and a large additional percentage most possibly suffer from mild discomfort, it is imperative that teachers of physical education give all high school and college girls an education in the correct use of exercises for the correction of this difficulty. Therefore no well-planned program of conditioning overlooks these exercises.

Since poor posture, particularly of the abdominal and lower back region, is so frequently a contributing factor to dysmen-

---

[3]John O. Haman: Exercises in Dysmenorrhea. Am. J. of Obst. & Gynec. 49:755 (June) 1945.

orrhea, the exercises given earlier in this chapter under the section on abdominal and lower back region should be taught to all girls. Furthermore, it is essential that exercises for strengthening the abdominal muscles be practiced daily by those who suffer from painful menstruation in order to develop proper support for the abdominal organs so that they will not sag downward and press upon the pelvic organs.

In addition, the following specific exercises will be found most helpful.

Exercise 1. Knee-Chest Position

If menstrual discomfort is due primarily to poor posture which allows the abdominal organs to sag and cause undue pressure on the pelvic organs or is due to a tipping backward of the uterus, assuming knee-chest position and remaining there for three to five minutes at a time often relieves discomfort. Assume this position by kneeling down and then placing the

Fig. 34. Knee-chest position.

chest as close to the floor as possible, turning the head to one side (Fig. 34). The arms may be placed in any position the individual finds comfortable. In this position gravity tends to pull the sagging abdominal organs upward and to pull forward and upward on a uterus which is tipped backward.

Exercise 2. Alternate Leg Extension From Knee-Chest Position

Sometimes extending first one leg and then the other backward-upward until it is in line with the trunk has a more beneficial effect than knee-chest position alone.

Exercise 3. Abdominal Pumping

This exercise, described in Chapter IV, is sometimes called the Mosher exercise after Dr. Celia Mosher, its originator. When properly executed it relieves pelvic congestion and increases pelvic circulation. When used as a daily conditioner it becomes useful as a preventive measure as well as relief for painful menstruation.

Exercise 4. Billig Exercises

These exercises were originated by Lieut. Comdr. Harvey E. Billig, Jr., M. C. (S) U.S.N.R. at the Harlan Shoemaker Clinic in Los Angeles[4] and have been used extensively and successfully in industry. Dr. Billig attributes many cases of painful menstruation "to a postural defect of contracted fascial ligamentous bands restricting posterior tilt excursion range of the pelvis in relation to the legs and back. At menstruation (with the change in ovarian hormone level) the ligaments further shorten, causing impingement with irritation of the peripheral spinal nerves passing through and by them . . . The impingement gives rise to symptoms of pain in the low back in the region of irritation or in the region of the distribution of these nerves . . . "[5]

For the correction of this condition Billig has invented two forms of "active progressive ligamentous stretching" exercises which if properly executed will sufficiently stretch the shortened pelvic fascial ligament bands to relieve menstrual pain.[6] One of these exercises is as follows:

a. Stand with left side toward a wall, heels and toes together, and just far enough away that the left forearm and hand may

---

[4]Estelle Gilman: Billig Dysmenorrhea Exercises. Aim, 5:18 (April) 1946.

[5]A. C. Dick, H. E. Billig, Jr. and Mrs. H. N. Macy: Menstrual Exercises—Absenteeism Decrease and Work Efficiency Increase. Indust. Med. 9:588-89 (September) 1943.

[6]These exercises were first successfully tried out in 1942 at the University of California, Los Angeles, and in the Los Angeles city schools. Beginning in July, 1942, they were given a ten-months trial at the Consolidated Vultee Aircraft Corporation of San Diego. The experiment eliminated 70,000 absentee hours due to dysmenorrhea. Eighty per cent of the former excuses because of menstrual cramps were cleared up, and there showed a definite efficiency increase of 70,000 hours in women who formerly suffered from menstrual discomfort but stayed on the job. It was found that from two to three months of work at these exercises were necessary for most moderate cases and from four to five months for severe cases. These facts are a matter of record in Dick, Billig, and Macy.[5]

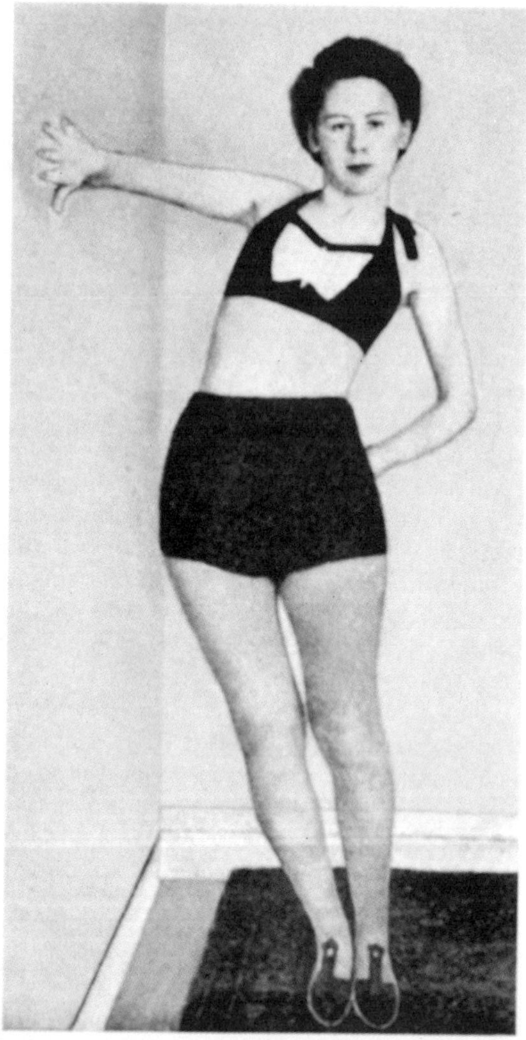

Fig. 35. Billig exercise, front view. (Courtesy Dr. Harvey E. Billig, The Billig Clinic, 536 South Hill Street, Los Angeles 14, California.)

be placed horizontally against the wall at shoulder level with the elbow directly opposite the shoulder joint. Place the heel of the right (or opposite) hand in the hollow which lies just to the rear of the prominence which may be felt over the right hip joint. Reduce the tilt of the pelvis by strongly contracting the

Fig. 36. Billig exercise, side view. (Courtesy of Dr. Harvey E. Billig, The Billig Clinic, 536 South Hill Street, Los Angeles 14, California.)

abdominal and gluteal (buttock) muscles and hold this position while forcing the knees straight, locking them in such a way that they seem to be forced backward. While forcibly maintaining this pelvic tilt and locked-knee position take a big breath, let it out, and then push the left hip toward the wall and slightly toward the thumb. Avoid any twisting of the body, sliding of the elbow up or down the wall, or any elevation of the left shoulder. The shoulders must be kept low and in line with the left upper arm which is perpendicular to the wall and must not be allowed to move forward. The knees must be kept straight and not allowed to bend (Figs. 35 and 36). The movement must be forced beyond the point of hurting if any progressive stretching is to occur. If the exercise is properly executed an individual is never able to push the hip far enough to actually touch the wall. Do three times with the left side toward the wall then repeat three times with the right side toward the wall.[7]

b. The original set of Billig exercises included a second form of stretching executed while facing the wall. Experience has proven, however, that this movement is so difficult to perform correctly that it is not recommended for general use.

When the Billig exercises are used to relieve painful menstruation they should be repeated three times a day from each of the two positions. This should be continued for two months or more until relief is obtained. Should relapse occur they should be done once a day during the week immediately preceding each period.[8]

MENSTRUAL FUNCTIONING AND EXERCISE

The amount and kind of exercise recommended for use at the menstrual period are discussed in Chapter IX.

---

[7]Harvey E. Billig, Jr.: The Release of Fascial Ligamentous Contractures in Physical Rehabilitation. Indust. Med. 14:270 (April) 1945.

[8]Haman: op. cit., p. 757.

CHAPTER VI

ACTIVITIES FOR DEVELOPING CARDIO-RESPIRATORY ENDURANCE

The role of exercise in relation to the heart, lungs, and circulatory system is explained in Chapter II. The use of running and rope jumping as two important ways to develop heart-lung (cardiorespiratory) endurance is suggested in Chapter IV. In the present chapter we will discuss the techniques of using these activities for this purpose and will also evaluate other activities in relation to their contribution to the development of cardiorespiratory endurance.

SUGGESTED LIST OF ACTIVITIES

All activities which strengthen muscles do not develop the heart-lung mechanism. Strength is developed in proportion to the intensity of the work done by the overload of intensity, while endurance is developed in proportion to the duration of the work done by the overload of duration. Progressive weight-lifting[1] is an excellent example of the overload of intensity without an overload of duration. If a man is able to lift a 25-lb. weight five times today and in four weeks is able to lift a 50-lb. weight five times, he has developed greater strength but has not necessarily increased his endurance. Or if a girl is able to do twenty Sit-Ups today with her arms folded on her chest, and a few weeks hence can do the same number (twenty) with her hands clasped behind her head, she has not increased her endurance because she has not increased the duration of the work, but she has increased her abdominal strength because it

[1]Progressive weight lifting as used by men is not advised for girls and women. Weight lifting increases abdominal pressure because the participants either hold their breath or breathe diaphragmatically. Any exercise which increases abdominal pressure is questionable for women.

is harder work (more intense work) for the abdominal muscles if the Sit-Ups are done with the hands behind the head than with them folded on the chest. When the main purpose of an activity is to develop cardiorespiratory (heart-lung) endurance the duration of the work is more important than the intensity. If a class is jumping rope in unison to a definite rhythm established by a drum or piano the girl who is able to continue jumping for the longest period of time has the greatest endurance. Those who can carry on the exercise but a short time in comparison with the majority of the group should be given special attention in physical education class for the development of endurance.

Cardiorespiratory endurance develops slowly so one needs to work at it for a long time. It is not developed to any great extent by gymnastic types of exercises alone unless they are strenuous and continued over a long enough period of time to tax the heart muscles and the lungs. Heart-lung endurance is increased only when the activity is prolonged beyond the first stages of breathlessness and discomfort to the "second wind" level.[2] The instructor should explain the meaning of second wind and use every means at her command to motivate pupils to continue an endurance activity until second wind has been experienced. Once pupils have learned that relief from the discomfort can be expected and once they have experienced this relief they develop an attitude of tolerance which aids materially in motivation toward the development of a high level of cardiorespiratory efficiency. The discussion of a few activities which can be used for this purpose follows.

Running

When running is used to develop cardiorespiratory endurance, distances long enough to prevent maximum speed throughout are more efficient than short sprints at full speed. Therefore, practice at a 200-yard run is better than work on a 50-yard dash and better still is practice at running for long stretches of time. Both the distance and the speed should be developed gradually, increasing either the speed or the distance from day to day but not both at the same time. The

---

[2]See Chapter II for explanation of "second wind."

ultimate aim in classwork when the running must be done around the gymnasium is to continue the run for a specified length of time at the fastest possible pace.

Running takes more energy than walking because it requires not only a higher speed of contraction of muscle but also requires more force since it lifts the body almost twice as much.[3] (See principles on pages 146 and 148.)

Techniques of running and methods of teaching are given later in this chapter.

Combination Hiking and Running

The development of endurance is only one of many objectives to be met in a single class period of thirty-five to forty-five minutes devoted to conditioning and body mechanics. Therefore, endurance should not take up more than its proportionate share of time. It can be supplemented by after school or week end hikes (See Chapter VII). A combination of walking and running can be used here with the length of the walk gradually decreased as the length of the run is increased. For example, start with a hike at the rate of four miles per hour. Then progress to an alternation of eight running steps and twenty-four walking steps, then eight running and sixteen walking, eight running and eight walking, sixteen running and eight walking, and twenty-four running and eight walking.

Another method is to keep a record of the time it takes each one of a group to traverse the distance of the hike, each individual competing not against the other members of the group but against her own previous performance on the same hike distance. The reward could be a prize to the one who reduced her time the most or a picnic lunch prepared by the losers who would be those who reduced their previous time record the least. This type of competition rewards even the slowest girl who brings up the rear, provided she improves her previous performance.

Still another way utilizes the motive of pressure from the group to improve individual performance. If there are several ways of equal distance of going from one point to another in

---

[3]Percy Dawson: The Physiology of Physical Education. Baltimore, Williams & Wilkins Co., 1935, p. 501.

a long hike, start one group on each path at the same time. The winning group is the one which manages to arrive in toto first. The winning group is decided not by the first girl to arrive but by the last. This has the advantage of stimulating the poorest to improvement but it fails to improve the performance of the best.

The correct technique for distance running is described later.

Obstacle Running

The use of an obstacle course is not as good as some other activities for developing endurance unless the course is longer than most, especially those which are set up indoors. When used they should put into practice the running, jumping, suspension, climbing, dodging, and other abilities built up or learned through other means. If the course is long enough and speed is stressed, then endurance may be developed.

If an obstacle course is used for girls, it should be one set up especially for their use and not one set up for boys. Many jumps and activities in a boy's course are too strenuous or unwise for girls. Girls who are menstruating should not participate in an obstacle run but may be used to assist at jumps, to time, or to judge.

Rope Jumping

Rope jumping has been of recognized value since ancient times and it should be used much more than it is in our school physical education programs. Agility is developed through rope jumping as awkward, heavy jumping gives way to easy light movements; rhythm is developed as jerky, angular movements of the rope give way to smooth, graceful, circular motions; coordination is developed as one learns variations and combinations of jump figures; and endurance is developed as one works at it and develops staying powers beyond the "second-wind" stage.

Sports and Dancing

Participation in competitive games and athletics and in all forms of dancing that give vigorous exercise is excellent for the development of cardiorespiratory endurance because of the intense motivation from within the individual which keeps her continuing the activity long enough to develop endurance.

In running, hiking, or rope jumping the gradually increasing dosage of activity is laid down by an instructor or a leader, but in sports and dancing the dosage is limited only by the individual and the intensity of her elation and desire to win or to perform. The intensity of this motivation from within usually carries an individual beyond the psychological limit of fatigue with less awareness of discomfort so that the activity is continued long enough to develop cardiorespiratory endurance.

Endurance is not necessarily developed by participation in all sports, not even in playing certain positions in some vigorous sports and not by certain forms of dancing. Sports and dancing, per se, do not develop endurance. It is only those sports and those positions on teams and those forms of dancing that require duration of vigorous exercise to the level of second wind and beyond, that develop heart-lung endurance.

One of the authors was a member of a group of thirty young American women studying under Niels Bukh in Ollerup, Denmark during the summer of 1929. She remembers with awe the high degree of endurance of the Danish young women who were studying there. One day a gymnastic demonstration was to be given at a place about fifty miles distant in which both the American and Danish groups had been asked to participate. One number on the program was an exhibition of folk dancing by the Danish girls. They had just finished a demonstration of about thirty minutes of as vigorous gymnastics as the author had ever seen when they started the folk dancing. This was continued for almost forty-five minutes without a single stop for rest except the slight pause necessitated by changing formations from dance to dance. Each dance was of an extremely vigorous type involving leaps, skips, polkas, and almost unbelievably high jumps into the air. When they finished the Danish girls were breathing deeply but not rapidly and gave the appearance of being able to continue for another hour. It left the American girls breathless with awe just to have watched such a performance.

TECHNIQUES OF RUNNING

The correct way to use the arms, legs, and body in running can be presented most easily in outline form. The important points are as follows:

1. Keep the body in a straight line but inclined slightly forward. (See basic principles for correct body mechanics, p. 147.) The degree of angle depends upon the speed desired. If the run is to be in the nature of a sprint, incline the body a bit farther forward than if the run is to be slower.

2. Hold the head up with the eyes to the front, lifting the chest in such a way as to give a feeling of leading with the chest.

3. Bend the arms at the elbows to approximately a right angle and keep in close to the sides; keep the forearms and loosely clenched fists pointing straight forward.

4. Swing the arms forward and backward, not diagonally across the chest but straight ahead and straight to the rear.

5. Swing the arms in opposition with the legs—right arm forward when left foot is forward, left arm forward with right foot.

6. Lift the knees a little higher than in walking but not too much.

7. To avoid the appearance of running stiff legged and to secure relaxation, swing the free lower leg up in back, but not too high.

8. Take long strides, swinging from the hip joint.

9. Keep the knees and ankles relaxed, "soft," to absorb the shock of body landing at each step. Landings should be light.

10. Keep the abdominal muscles taut so that the viscera will not be jarred by the swift transfer of weight from one foot to the other.

11. Use the feet in a straight forward and backward direction with no toeing out.

12. Land on the ball of the foot. The heel should not strike the floor.

The correct technique of running for longer distances out of doors is the same except that the forward inclination of the body is slightly less (but still forward 5 to 10 degrees), the arms are not bent quite as far, the heel is lowered closer to the ground and may even touch momentarily, and the stride is easy and relaxed. Breathing should be through the nose and mouth.

Warning

Never allow a class or an individual to start running at full speed without warming up first. This warm-up could be a preliminary run at an easy pace for at least one minute. Have pupils finish all runs with a short walk before going to the showers.

Special Advice on Teaching Running as a Class Exercise

The teacher needs to be on the alert if her class is to derive real benefit from the running. It is not sufficient to tell a class to run and then leave them to make whatever they desire out of the exercise; some jogging along slowly get in the way of those behind, while others driving themselves with furious spurts run down everyone in front of them; and some running a few rounds drop out at the first symptom of breathlessness, while others hang on too long.

A teacher who really teaches running to her pupils has control of the class every moment, guarding all for safety measures, setting rhythm, admonishing individuals, watching each for signs of undue exertion (greatly heightened color or undue pallor, look of strain on the face, undue breathlessness), and calling on some to ease up or drop out and on others to keep trying or to work harder.

The authors have seen some otherwise good physical education teachers start classes running at the close of a class period and then disappear leaving the pupils to run as little or as much as they like; or if not leaving the floor, the teacher has turned to converse with visitors or class members who are not running, paying no attention to this feature of the class activity. If running is to be either an educational or a physical developmental part of a class period or both, as it should be, it requires the undivided attention of the teacher. Some definite suggestions for organizing practice in running are outlined as follows:

1. Emphasize form in running before speed.

2. Demonstrate good running form, explaining the principles involved.

3. Arrange the class in two groups at the two ends of the same side of the room and line up each group into rows with six to a row. Have the first row of Group A do a practice slow run to the vacant corner of the room diagonally opposite them. The moment they are past the center of the room the first row of Group B runs to its diagonally opposite corner. As soon as it passes the center of the room the second row of Group A follows its first row, and so on until all the class, six at a time, have had a trial run.

4. Call for criticisms of this trial run. Admonish and demonstrate correct form once more.

5. Repeat Numbers 3 and 4 several times.

6. Have the class run slowly in circles about the room. If the class is large they will need more than one circle so that each pupil will have plenty of space so as not to crowd her neighbors. Form a second circle within the outer circle.

7. Gradually step up the tempo of the run from a jogging step to a fast run.

8. Once a season have a cross-country run, if possible.

Students should be taught that it favors heart recovery after a period of running to do some light arm or leg exercise or both, rather than to stand still, sit down, or even lie down. This is provided for in the conditioning series by walking long enough after the initial run to "open order" and by the nature of the first few exercises.

TECHNIQUES OF ROPE JUMPING

Rope jumping is vigorous and interesting, and if pupils are constantly stimulated to learn new techniques, it can hold their interest over a long period of time.[4]

Equipment

The equipment is inexpensive. If necessary each pupil can bring a length of rope from home. There should be a sufficient number of ropes for the largest number of jumpers the floor

[4]Alice Marble claims to have achieved her unusual tennis footwork through rope-jumping skills which she acquired the hard way—in fact, the only way one does acquire such skills, namely, by dint of long, hard hours of practice.

will accommodate at one time. The ropes should measure from 7 feet to 8 feet 6 inches in length and should be of a quality that will not twist on itself.

The ropes should be kept on a rack or in an easily accessible cupboard with the different lengths hung together on the correctly marked hook or peg. They should be available for students to use whenever they have free time to work with them.

Fundamentals of Rope Jumping

In all variations of rope jumping the emphasis should be placed upon reducing the amount of total arm action with motion in the shoulder joint, and increasing the amount of flexible wrist action with the arms kept fairly close to the sides. Action of the feet and legs should be reduced to the minimum which will clear the rope. The individual who jumps eight inches into the air to clear a half-inch rope is wasting energy and learning to do a movement inefficiently. When she learns to jump just high enough to clear the rope, she has learned efficiency in action. Swinging the legs vigorously backward and forward with excessive knee action when the feet are used alternately is also inefficient and should be discouraged in favor of maximum ankle action with a minimum of knee action.

Motivation

If heart-lung endurance is to be developed pupils must be motivated to continue jumping beyond the stage of "second wind." For this the motive of competition is helpful. For example, a group may start jumping on signal and in unison and continue for as long as possible, each girl sitting down in place when she can no longer continue, and those who are still jumping at the end of a certain time are declared to have the best cardiorespiratory endurance. Another method is to set a time limit, start the group jumping upon signal, each individual attempting to make as many jumps as possible in the time set. Each one counts the number of jumps she is able to do and, upon the signal to stop, the girls who were able to maintain the fastest pace for the longest time, thereby accumulating the most jumps, are declared the winners. Just as the 200-yard run at a fast pace is better for developing endurance than the 50-yard dash at the same pace, so the maximum number of jumps possible in three or four minutes is better than one minute for developing endurance.

Variations of Rope Jumping

There are many combinations and variations of rope jumping which can be used which are challenging and fun to do. Some involve individual jumping, some group jumping, while others have individuals turning small ropes inside large ropes swung by other pupils. Individual jumping is preferable when the purpose is the development of cardiorespiratory endurance or the maximum amount of exercise possible in a given length of time.

Some individual movements to be done with small ropes are outlined below.

Variations In Which The Performer Jumps On Both Feet Simultaneously

1. One-count jump forward. Jump over the rope once on each forward circle of the rope.
2. Two-count jump forward. On each forward circle of the rope jump twice, alternating a jump over the rope with a jump before the next forward circle.
3. Crossing the feet. Jump only once, over the rope, on each circle of the rope forward but cross the feet each time, right foot in front of left on the first jump, then left foot in front of right, and so on, reversing the position of the feet on each jump.
4. One-count jump forward with double rope circle. The rope makes two complete circles forward and under the feet twice on a single jump into the air.
5. Two-count jump backward. The rope circles backward, the performer jumping twice on each circle, once over the rope, once before the next circle of the rope.
6. One-count jump backward. The rope circles backward, the performer jumping once on each circle, over the rope.
7. Two-count jump forward with crossed arms. The performer crosses the arms in front and swings the rope forward by action of the right wrist on the left side of the body and the left wrist on the right side. Jump twice on each circle of the rope, once over the rope, and once before the next circle. Be-

ginners find it easier if the elbows are kept in close contact in front of the body and they learn to swing the rope mostly by wrist action.

8. Two-count jump backward with crossed arms. Action the same as Number 7 except the rope swings backward. Beginners must learn to keep elbow contact. Strangely enough, this seems to be easier than the forward jump with crossed arms and should probably be taught first.

9. Two-count jump forward alternating natural and crossed arm position. The rope circles forward alternating one circle with the arms in their natural position with one circle crossed over to the opposite side. The feet make two jumps on each circle, once over the rope, once before the next circle.

10. Two-count jump backward alternating natural and crossed arm position. Action identical with Number 9 except the rope circles backward.

Variations In Which The Performer Jumps On One Foot At a Time

11. Two-count jump forward alternating feet. Jump twice on each circle of the rope forward, once over the rope on the right (left) foot, once before the next circle on the right (left) foot. Then repeat with the left (right) foot and continue to alternate feet.

12. Two-count jump backward alternating feet. Same as Number 11 except that the rope circles backward.

13. One-count jump forward alternating feet. Jump only once on each forward circle of the rope, over the rope, first on the right (left) foot, then on the left (right) and continue to alternate feet on each jump.

14. One-count jump backward alternating feet. Same as Number 13 except that the rope circles backward.

15. Running forward progressing down the room, jumping over rope on every other step.

16. Running forward progressing down the room, jumping over the rope on every step.

17. Single one-count jump forward. Hop over the rope with the same foot continuously for a number of times before changing feet. Any combination of times may be used, 16 on each, 8 on each, or 8, 4, 2, 1 alternately, and so on.

18. Single one-count jump backward. Same as 17 but the rope swings backward.

19. Single two-count jump forward. Hop twice on the same foot on each circle of the rope forward, once over the rope, once before the next circle, for any number of times, then repeat on the other foot.

20. Single two-count jump backward. Same as Number 19 except that the rope swings backward.

The above are only a few of the infinite number of variations possible. It often proves interesting to let the class suggest or work out new combinations.

Warning

Since there is a question as to the advisability of girls doing any activity involving jumping at the menstrual period, it is advisable to have them omit rope jumping activity at this time. While the other members of the class are jumping these girls might well be running.

CHAPTER VII

EFFECTIVE USE OF THE BODY FOR EVERYDAY LIVING

When you go to the legitimate theater or a movie do you ever try to analyze the body movements of the leading lady to find out just what it is that she does with her body that makes her movements so pleasing? She doesn't clump or waddle along, fall disjointedly into a chair or hitch herself up out of one, and sit or stand awkwardly. What is it that she does with her body that sets her apart from the usual run of women? Whatever she does, it most probably did not come to her naturally. No doubt she has been carefully coached and has painstakingly practiced the art of handling her body effectively and pleasingly. What she has accomplished others can accomplish.

Some (the few) are endowed by nature with a fine physique and, of these, a few maintain good health conditions and achieve good body mechanics with but little effort. However, the great run of mankind must be educated about and drilled in the proper use of the body.

Since the effective use of the body in all of life's activities has such an important relation to one's health, success in, and enjoyment of life it is difficult to understand why this part of a child's education is so universally neglected in both the home and the school. Walter Pitkin criticizes modern education in that it does not teach the common man important facts and skills for everyday living, such as how to give oneself proper nutrition, how to work efficiently avoiding exhaustion, how to rest and play effectively and how to walk, run, and handle the body in other everyday activities in a capable manner.[1] His challenge to education is particularly a challenge to the phys-

[1]Walter B. Pitkin: More Power to You. New York, Simon & Schuster, Inc., 1933, p. 21. Adapted with permission of copyright owners.

ical education departments of our schools, but it is also a challenge to the administrators to support these teachers in their efforts. The part which the physical education teacher should play in such an educational program is discussed in this chapter. The information given below and the suggested skills practices should be woven into a course which should be given to all school children. How to organize this material into a unified course is discussed in Chapter XII.

In the preceding three chapters there have been described and explained exercises to build strength and endurance and for general body conditioning. In this chapter are presented the techniques of effective use of the body after strength and endurance have been developed, also techniques for more efficient living as one keeps in condition.

The manner in which a person uses his body as he stands, sits, walks, stoops, bends over, reaches, lifts, pulls, pushes, carries, sleeps or does any other thing determines whether he works and plays and lives effectively with a minimum of fatigue and strain. Good use of the body means getting the best results with the least effort. There is a right way to do everything, and the right way is the efficient way, the energy-saving way. The wrong way causes strains, drains, fatigue, irritations, and frustrations, and it is wasteful of time and energy. Work that is done at maximum efficiency is that work which is done with the least possible expenditure of energy.

VALUES OF THE EFFECTIVE USE OF THE BODY

Spiritual Values

The self-respect and self-confidence which beget good posture are in turn strengthened by good posture. The slouching attitudes of the ineffective, the slinking gait of the incompetent, the drooping head and chest and round shoulders of the inept, the too straight back of the narrow minded, the too-far-back shoulders and head of the arrogant, the too chesty attitude of the overconfident, are all familiar signposts for character reading. As these attitudes are the outward sign of ineffectual living, so efforts at living effectively may be strengthened by a direct attack on posture itself. Who has not experienced the effect on one's determination of squaring the shoulders, lifting the head, and throwing out the chest, or the effect on one's own self-appraisal of standing as erect and tall as possible?

Darwin in his Expression of "The Emotions in Man and Animals" spoke of the intimate relation that exists between emotions and their outward manifestation and of how an emotion is intensified by giving free expression to its outward signs. For example, it intensifies one's actual courage to act and appear courageous. William James in his "Psychology" also discussed this effect of posture and attitudes on the inner emotions. He called attention to the fact that flight from fear increases panic, and that a moping posture and dismal voice increase melancholy feelings. He advised as one step in educating for correct mental attitudes the training in assuming the outward symbols of those mental attitudes which we wish to cultivate. Therefore, since good posture gives the outward impression of buoyancy and nobility of spirit, of courage and physical vigor, of cheerfulness and hope, it is important that every child be trained in the techniques of good posture if for no other reason than that his buoyancy and nobility of spirit, his courage and physical vigor, his cheerfulness and hopeful outlook on life may be intensified.

To teach every child to attain the outward signs of inner spiritual effectiveness is a task which the school and the home should undertake as one step in the spiritual development of the child. This training in good posture cannot be separated from character education, and character education is not complete without educating the body in the art of good posture as an outward symbol of inward values.

Social Values

What means more to a woman socially than the common appraisal: "She carries herself like a queen?" The battle of winning recognition socially is half won with good posture, and it is more than half lost with poor posture. Society in general accepts a person at his own appraisal, and his posture speaks louder than words—in fact it is almost entirely posture that gives the first impression of a stranger. Is that first impression one of self-confidence, self-respect, good health, vigor, and effectiveness? It is quite sure to be if the posture is good.

A study of the posture of famous people shows that the world's eminent people are people of good posture. A study of artists' conceptions of social leaders through the ages shows persons of magnificent bearing. In the theater, actors express success and unusual leadership qualities through the postures they assume. In fact the successful actor goes through years of tedious training in good posture and good body mechanics if he is not fortunate enough to be endowed with them as a natural heritage. The success stories of countless actors, singers, public speakers, and world leaders are stories of meticulous attention to body conditioning, posture, and personal bearing in general.

Social Values as Portrayed in Art

When an artist wishes to convey the idea of nobility of soul, inspiration, superiority, gladness, hope, or intelligence, he depicts a person of majestic carriage. Children should become familiar with good posture as shown in art. Classwork on posture gives the physical education teacher excellent opportunity to correlate physical education classwork with the study of the great masters of art. Many of the world's famous pictures offer excellent opportunity to call pupils' attention to the good posture of famous persons. A special display by the school's art department of reproductions of such pictures would materially aid the physical education teacher when she is working on this subject. Any school that possesses even a small collection of reproductions of the work of famous painters and sculptors will be sure to have in the collection pictures of persons of nobility and courage; these are the pictures that will show good posture, for artists throughout the ages have recognized this physical manifestation of the fine spirit. A suggested list of excellent reproductions for such a display is given in Chapter XI.

Economic Values

Employers are reluctant to employ persons of poor posture since they are fearful of their health condition and of their seeming lack of stamina. And all too frequently their fears are well grounded. An applicant's seeming lack of self-confidence or his seeming overconfidence, depending on the type of poor posture, also causes employers to hesitate. On the other hand, a person of good posture and confident bearing sells himself readily as far as outward personal qualifications are concerned.

Health Values

Good posture of itself does not give good health, but it has profound effect upon the quality of life one leads. Maintaining the various segments of the body in correct relation to each other directly affects the working capacity of the vital organs which are supported by the skeletal structures. For example, a cramped chest and forward shoulders may cause cramped position of the heart and lungs with subsequent loss of effective functioning; forward head with its constant pulling on the neck muscles causes headaches and strain in the neck region; relaxed abdominal walls permit relaxation of abdominal organs from their best working position; incorrect pelvic tilt causes lower back strain, and incorrect weight bearing on the feet produces numerous aches and pains not only in the feet but in the legs and back as well. The best functioning of the body for all-round good health calls for good posture as one important consideration.

BASIC PRINCIPLES FOR CORRECT BODY MECHANICS

There are postures and movements which are common to all tasks and occupations and there are basic principles which characterize their efficient performance. Therefore understanding of these fundamental principles is necessary in order to determine the most effective way of performing a task. Taking into consideration both the structure and the functioning of the human body and the laws of motion and the principles of muscle action, the authors have formulated a set of principles that seem to be basic to efficient movement.

As far as mechanics are concerned, the human body presents a problem of weights, levers, supports, and devices for producing motion. The body mass structure is the weights; the joints, the levers; the bones, the support; and the muscles, the devices for producing motion. The study of body mechanics is therefore the task of determining how the body weights can be handled with the least strain and least expenditure of energy so as to maintain stability when at rest or in motion. These weights are handled best when the basic principles of correct mechanics are applied.

Basic Principles[2]

1. Correct body mechanics are those that produce the best results with the least effort.

   a. The least effort is required when those muscles selected to do the work are the strongest of those which can perform the task. (The leg muscles are stronger than the back muscles and therefore they, instead of the back muscles, should do heavy lifting.)

   b. "A muscle can exert its greatest force when it is fully extended, and as it shortens, its force diminishes."[3] (In other words, to get strong work out of a muscle, first put it on the stretch; the less the force required, the less the amount of stretch that is needed.)

   c. Low speeds of contraction of muscle require less energy than high speeds.[4]

   d. Work done with vigor is more economical of body energy than that performed laxly.[5]

   e. The greater the coordination of the muscles, the less the amount of work required of them.[6]

2. Opposing forces are constantly at work on the body[7] so that there is ever present the task of holding these opposing forces in equilibrium.

---

[2]The principles set forth here should be presented and explained to the pupils and, as classwork proceeds on the techniques of effective use of the body, these principles should be referred to constantly. It would be well to have them boiled down to the most essential facts and these posted for pupils' out-of-class reference and study.

[3]Wilber Bowen and R. Tait McKenzie: Applied Anatomy and Kinesiology. 4th ed. Philadelphia, Lea & Febiger, 1928, p. 26.

[4]"The greater the speed of contraction, the more energy is required to overcome the viscosity of the muscles and therefore the more energy is lost in heat" from Percy Dawson: The Physiology of Physical Education. Baltimore, Williams & Wilkins Co., 1935, p. 213.

[5]Less carbon dioxide is eliminated in vigorous work (Ibid., p. 214).

[6]Ibid., p. 215.

[7]Newton's Third Law of Motion: To every force action there is an equal and opposite reaction.

a. When opposing forces which act upon a body are equal, stability is maintained in that body.

b. When opposing forces are not equal, stability is disturbed until additional force is available to supplement the lesser force. (If the requirements for additional force are too great, strain results.)

3. An object at rest tends to remain at rest and an object moving tends to continue moving in the same direction and at the same speed unless some external force intervenes.[8]

a. In moving, the body must overcome inertia.

b. Once a movement is started it takes less effort to maintain a given speed than to change the speed (because of momentum).

c. The more the use of momentum the less the effort required.

4. Gravity acts upon the body in a vertical downward pull.

a. The stability of the body is most efficiently maintained when the body segments are centered over each other and the base of support and along the line of the pull of gravity.

b. The greater the angle of inclination of the body or any segment from the vertical position, the greater the muscular effort required to maintain balance.

(1) When body balance is disturbed a broader base of support makes the maintenance of balance easier.

5. When the body is to be put in motion, force should be applied through the center of gravity[9] in the direction of the desired movement.

a. The circumstances are the most favorable for correct mechanics in movement if the body segments are maintained in correct alignment in relation to each other.

b. When the body is moved vertically upward or lowered vertically downward, less effort is required if the center of gravity is kept directly above the base of support.

---

[8]Newton's First Law of Motion.

[9]According to Lipovetz the average center of gravity for women is 54 per cent of height—or 2.3 inches above the ileum. Lipovetz, F. J.: Medical Physical Education. Minneapolis, Burgess Publishing Co., 1947, p. 20.

c. When the body is moved in a forward-upward direction (as in running, broad jumping, and so forth) it is more efficient if the center of gravity is in advance or ahead of the base of support.

6. When lifting, supporting, or carrying an object, that object becomes a part of the body weight in so far as forces that maintain balance are concerned.

a. The closer the load is kept to the center of body gravity and the more directly over the base of support the less the expenditure of energy.

b. Moderate weights are more economical of energy than extremely light or extremely heavy ones. (A moderate weight is one which is about half of what a muscle can lift.)[10]

c. The work of handling a load is performed with the least expenditure of energy if the muscles that are to do the work get set before contacting the load. (In other words, prepare for the load.)

d. The greater the distance of the lift of either body or object or both, the greater the amount of work.

(1) When stooping to a load, the body should be lowered only so far as is necessary to reach the desired object and at the same time to put the leg muscles on sufficient stretch for the work to be done.

7. In moving a load less energy is expended if the force is applied through the center of gravity of the load in the direction of the desired movement.

8. It requires many repetitions of a desired action before the reflex arcs of the nervous system are effectively "conditioned" and the action is done without conscious control. (If habits of efficient body mechanics are to be formed, each repetition of the action must be done in the correct manner.)

The Teacher's Responsibility Toward Basic Understanding

The teacher of a course in body mechanics should be familiar with the above principles and their application and should

[10]Industrial research has established the opinion that the average man can handle a shovel load of 21 pounds to the best mechanical advantage; therefore the standard shovel is built for that load. (Bowen and McKenzie: op. cit., p. 27.)

lead her pupils to such an understanding as is consistent with their age-level capacity. Beyond this she should guide pupils into the experiences of figuring out for themselves the most efficient way to handle the body. Since it is out of the question to expect pupils to have studied anatomy, physiology, physics, kinesiology, chemistry, and psychology (all of which studies enter into an understanding of how the body is constructed, how it functions, and how it moves) before learning the techniques of efficient body management, the teacher herself should have these understandings and should be prepared to give to the pupils the necessary fundamental facts for their intelligent cooperation in the classwork.[11]

Needless to say, a child learns to walk, stand, run, and handle its body in all of life's situations long before it has the intelligence (let alone the desire) to study any of the above mentioned subjects, nonetheless he is able with the guidance of good teaching to acquire good body mechanics; for the young child understanding must come later. This situation exists in varying degrees in teaching any form of skills at any age level. The teacher, whether parent or other leader, will give better guidance if her understanding of fundamentals is based on scientific knowledge, but the child must be guided without waiting to develop its own capacity to understand all of these facts. The wise teacher will temper to the particular age of her group the amount of scientific facts she will give them. These fundamentals of scientific facts should not take up an undue amount of class time. The teacher should sift out the most important facts for class presentation yet at the same time she should conserve class time to work at the techniques of the activities involved.

Teaching the important facts and instructing in fundamental skills, alone, are not enough! There must also be much practice of the skills on the part of the pupils. Every time a movement is performed in the correct way one more step is made toward developing that correct movement into a habit. If

---

[11]Mabel Elsworth Todd has made a searching study of the basic principles of body mechanics as is recorded in her book, The Thinking Body (Paul B. Hoeber, Inc., 1937). The teacher would do well to use it not only as a "refresher" on structure and functioning of the body but also for a basic understanding of the laws governing body mechanics.

correct posture and correct movement are practiced by conscious control over a long enough period of time, the reflex arcs of the nervous system eventually become "conditioned" for correct posture and correct movement. In other words, habits of correct body mechanics have been formed and conscious control is no longer required.

The ultimate aim in teaching body mechanics, then, is, first, to give students the facts or principles which are fundamental to correct movement; second, to teach them to analyze postures and movement in terms of these principles; and, third, to provide an opportunity for sufficient practice of some of the more common everyday activities so that habits of efficient movement will be formed.

STANDING POSTURE

Most girls and women are interested in how they look in a bathing suit or evening clothes. They know that a good figure adds materially to one's attractiveness and that a poor figure can give a feeling of inferiority. Good posture goes a long way toward giving that fine appearance that helps land a position or win social recognition. What girl doesn't desire to wear her clothes well? Fine clothes cannot offset poor posture, but good posture transcends poor clothes. Posture expresses personality even more than do voice and clothes. In most of life's situations people have opportunity to look a stranger over before hearing her speak so that it is posture and bearing that give that all important first impression.

The ability to maintain a correct standing position is basic to handling the body correctly in all other activities.[12] Since we live with our bodies in motion much more than in static positions, it is important to make pupils understand that correct posture is a moving posture as well as a "standing still" one. Learning to balance the body segments one above the other and keeping them in line with the pull of gravity is the keystone of posture training. In the best mechanical position the body parts support each other; "we" do not have to hold the body up. However, if the position of the body is not mechanically

[12]Note the care by highly skilled performers (circus performers, dancers, ice skaters, gymnasts) to assume a correct standing position at the moment just preceding the "take off."

correct, the muscles must work to hold up parts against the pull of gravity and this means wasted energy and strains. At best, man's upright position over a small base of support is a difficult one. Learning to handle this mechanical problem effectively is a real victory which frees one for attention to other worth-while ventures in living.

The Anatomy of Good Posture

The "consistent pattern of alignment", of the bony framework of the body sets the pattern of posture (Fig. 37). The bones are held together by ligaments. The junction of two bones makes a joint. Muscles are attached to bones by tendons, each muscle being attached to two different bones so that when the muscle contracts it tends to draw closer together the parts of the two bones to which the muscle is attached, the movement taking place in the joint between the two bones. For every joint there are opposing sets of muscles, one producing a given action, the other restoring to the original position. For example, one group of muscles which is attached to both the forearm and upper arm will, when contracted, bend the elbow, bringing the two parts of the arm together, while the opposing group of muscles is so attached to the same two bones that when it contracts it will "unbend" the elbow, returning the two bones to a straight line with each other. When opposing muscles pull with the same force at the same time no movement results, but the joint is well supported. If both groups of opposing muscles relax at the same time there also is no movement but the joint is not supported. When a given movement is desired one group of muscles relaxes as the other contracts; when the opposite movement is desired the muscles that were relaxed contract and those that were contracted relax. The reader would do well to study Figures 1 and 2.

The positions which the various segments of the body hold in relation to each other for habitual position determine one's posture and these positions are determined by the musculature of the body. This statement is sufficient for our present purpose. We do not need to enter here into a discussion of postural difficulties of shortened or lengthened tendons and of abnormal musculature and bony structure. For the great majority of people muscle education is the whole story of posture training,

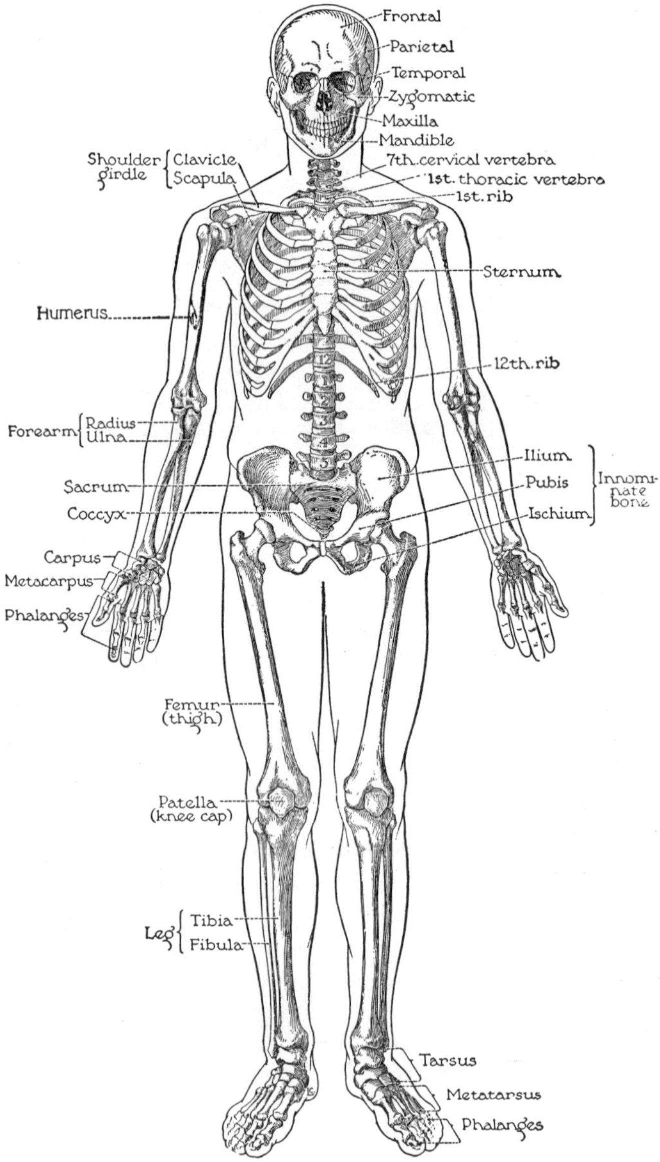

Fig. 37. Anterior view of the human skeleton.
(Millard and King: Human Anatomy and Physiology.)

and since good posture is not the natural inheritance of the general run of mankind it is evident that training in this direction should be given to every child, and that those conditions which lead to good muscle tone must be sought for every child. Such conditions involve proper nutrition, sufficient rest, and large daily doses of exercise.

Body Alignment

Careful attention is usually given to man-made machines to keep their parts in correct alignment for proper functioning, but the human body, the most wonderful mechanical machine of all, is too frequently neglected and must do its work with its parts out of proper alignment. There is a certain relation of the various body segments to each other which places them in a position in which the body as a whole functions at its best. This particular relationship or alignment is known as "correct" or "good posture." When the person is standing in this correct relationship the various body segments are in line one directly above the other so that they support each other along the line of the pull of gravity (Fig. 38). This line of pull is called the long axis of the body. An imaginary line drawn from side view through the top of the head, just back of the ear, through the center of the shoulder joint, just back of the center of the hip joint, just in front of the center of the knee joint, through the center of the ankle joint, is a straight line if the body-segment alignment is correct.[13] Any noticeable position of one of these parts either forward or backward produces a break in this imaginary line, and where the break occurs there will be strain on the body balance resulting in unneccessary fatigue and poor functioning. Figure 38 shows how this imaginary line is broken by poor alignment of body segments.

Ability to maintain the body segments in correct alignment depends primarily upon the art of keeping the body segments balanced one above the other, centered over the base of sup-

---

[13]There is difference of opinion among authorities as to the exact location of this line of gravity pull so that one cannot go far astray if this imaginary line falls slightly forward or backward of the particular locations mentioned above. The authors accept the research findings of the Webers which designate the two points just in back of the hip joint and just in front of the knee joint, as reported by Percy Dawson: op. cit., p. 458.

port and in line with the pull of gravity; beyond this it also depends upon the strength and endurance of the skeletal muscles and the skill and will to use these muscles efficiently.

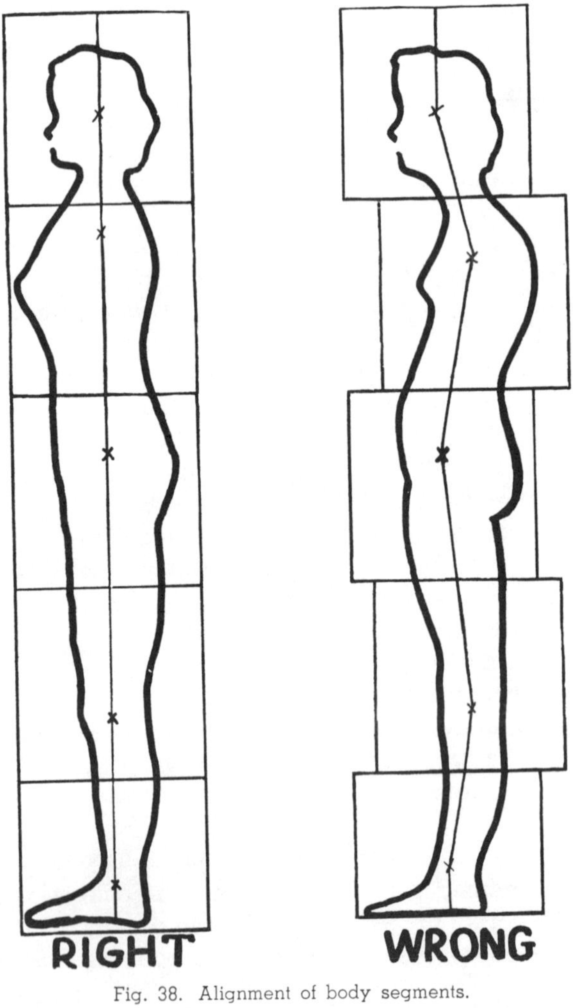

Fig. 38. Alignment of body segments.

Kinesthetic Awareness

Kinesthetic awareness in relation to posture is an awareness of the body masses and how they are aligned and of the tensions which result from poor alignment. Development of this awareness is a first step in posture training. In developing kinesthetic awareness the teacher should start with relaxation and the recognition of tension and lack of tension. She should help the pupil to interpret this kinesthetic awareness in relation to muscle sensations which accompany movement. Following are a few suggestions for teaching aids in developing kinesthetic awareness:

1. Mental concepts, such as making the arms feel heavy and long or making the shoulders feel wide as though reaching far out sideways.

2. Use of the hands to help develop awareness of tension or lack of tension. For example, place a hand on the abdomen or on the buttock muscles, tense them and note the tensing sensation. Then strive to get the same "feel" without laying the hands on the parts.

3. Use of a mirror to associate the picture of a correct movement with the "feel." Students soon learn to note their own mistakes.

4. Use of the concept "as if." For example, stand "as if you have a tail and are tucking it in" or "as if you are squeezing through a tight space."

5. Closing the eyes and concentrating on the "feel" in the muscles and joints while they are in various positions.

Poor Posture

As is so frequently said, bad posture not only looks bad, but it is bad—bad for one's appearance, bad for one's efforts at success in life, and, in cases of extremely bad posture, it is bad for one's health. It puts a strain on the lower back causing backache; it makes the menstrual cycle difficult; it causes general fatigue and lowered muscle tone. In extreme cases, it may hamper heart and lung mechanism and interfere with normal digestive functioning. All these ill effects react unfavorably on both the mind and the spirit, producing a poor outlook on life. As one's body sags, so one's spirits sag. "Build thee more stately mansions" is an obligation of the body as well as of the soul for each affects the other immeasurably.

Indications of bad posture are forward head, sunken chest, round shoulders, hollow back, protruding abdomen, protruding buttocks, locked knees, and weight thrown on the inner borders of the feet. Penalties of poor posture are "dowager's hump," neckache, backache, double chin, winged scapulae, fallen arches, "middle age spread" and "spare tires" around the waist.

Good Posture

The body is the earthly dwelling place of the soul. It should be made as beautiful as possible with the material with which nature has endowed us. Our bearing should be worthy of a fine spirit: It should reflect self-respect, self-reliance, dignity, courage, and strength of will. Heredity determines individual differences such as whether we shall be short or tall, lanky or heavy set, large boned or small boned, but within these limitations each can seek to make as mechanically perfect as possible the body type he has inherited.

A Twenty-Four Hour Job

It is not just a few minutes of exercise each day but the way a person handles his body twenty-four hours a day that determines his posture. "Pretty is as pretty does" most certainly applies to posture. The way we do such things as reach, lift, carry things, stand, sit, walk, and run decides whether we go through life under constant strain or with vim and zest and alertness.

Handicaps To Good Posture

Handicaps to good posture are numerous. The following list covers the major items:

1. Bed covers too tight over feet
2. Chronic fatigue
3. Clothing too heavy on shoulders
4. Defective vision
5. Defective hearing
6. Discouragement
7. Frequent illnesses
8. Heels too high
9. High heels worn too constantly
10. Lack of self confidence
11. Poor nutrition
12. Shoes run over at heels
13. Soft and sagging bed
14. Study tables too low
15. Too constant one-sided use of the body
16. Too little exercise
17. Too little free play
18. Too little rest
19. Underweight
20. Weak musculature
21. Work benches or seats poorly adjusted

Cornerstones To Good Posture

Of prime importance to good posture are strength of muscle, endurance of the heart-lung mechanism, and coordination of muscle and mind. These cornerstones must be laid first and then on these are built the skill of body control. Some people do not have good carriage because some muscles are overdeveloped and some are underdeveloped. Some people rely on proper diet alone to give them a good figure. Others rely wholly on exercise alone but it takes proper diet plus proper and regular exercise plus constant effective use of the body parts in relation to each other to produce and maintain good posture.

The Mechanics of Good Posture

The one basic rule which applies to the mechanics of good posture is that the various body segments should support each other in vertical alignment and directly over the base of support. All details of the mastery of correct posture are but amplifications of this basic rule. When one body segment is moved out of line with the other body segments which themselves are centered over the base of support and in perpendicular alignment with the pull of gravity, then additional force must be exerted to maintain the balance against the "off weight." This additional force represents not only wasted energy but also unequal pull on muscles and ligaments, producing strains. (Fig. 39.)

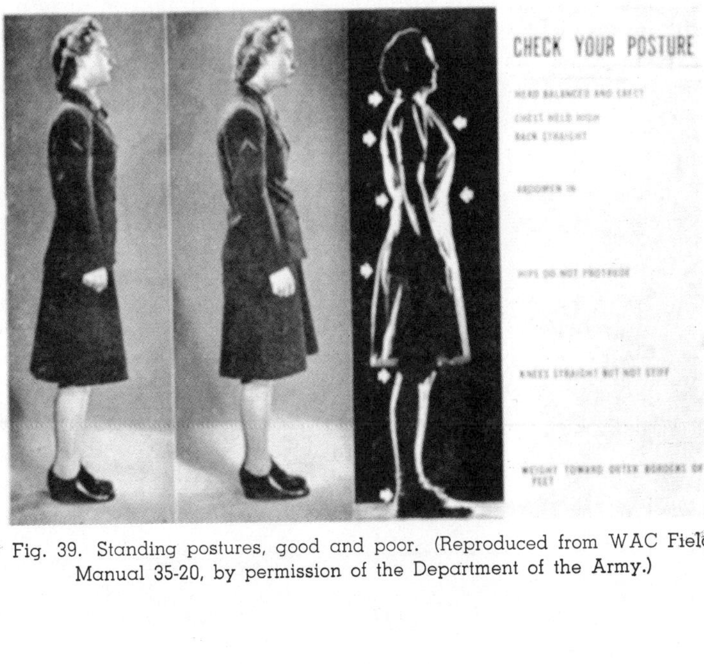

Fig. 39. Standing postures, good and poor. (Reproduced from WAC Field Manual 35-20, by permission of the Department of the Army.)

Good posture means not only body segments in proper alignment, but also muscles working in good coordination, on the alert yet relaxed. It is much easier to acquire a good posture when young and supple and then keep it by constant training and watchfulness than to try to attain it when older after the neglect of youth has caused damage which must be undone before reconstruction work begins.

Although all posture training should proceed from the general to the specific and back to the general, as is suggested in the class techniques offered later in this chapter, the various specifics will be discussed in the material that follows. The teacher should realize, however, that practice on specifics alone will not give coordinations for good posture of the body as a whole.

The Head in Relation To Good Posture

The head is a heavy segment of the body and is attached to the segment under it by a very small and pliable connecting rod (the neck). Therefore it should be poised and balanced just so or it will put great strain on the neck and shoulders. The neck muscles control the head position, and they must be well trained for this delicate balancing act with its heavy load. The chin should not be thrust forward but should be drawn in, although not tensely giving the effect of a double chin; it should be elevated to such a degree that the eyes can look out straight ahead at eye level without need of adjustment. (See Fig. 40.)

Since most people are inclined to keep the head down too much or the chin thrust too far forward the catch words for good head posture are: "Eyes front" and "Chin in and up."

The Shoulders in Relation to Good Posture

The habitual position of the shoulders reveals much about personality. The forward shoulders suggest subservience; the tense and stiff shoulders indicate stubborness, hence the saying "He is so stiff backed;" the too-far-back shoulders bespeak the cocksure; and the twitching, twisting shoulders announce the erratic person. For good posture the shoulders are held as wide as possible with the tip of each pointing directly to the side. They also are relaxed, not hunched upward. It is poor

"The Column of Youth"

Fig. 40. Sculpture by R. Tait McKenzie which is enshrined in the National Education Association Building in Washington as a memorial to the sculptor. (Photograph reproduced by courtesy of The American Association for Health, Physical Education, and Recreation.)

shoulder posture to have the shoulder blades (scapulae) forced too close together along the spine or to have them tipped away from each other with the lower angles protruding, the position known as "winged scapulae." The position of the shoulders should be such that the scapula rides over the ribs at the rear, holding a position parallel to them.

For good shoulder posture never say, "Throw back your shoulders," but instead say, "Shoulders down—make them wide." Think of the shoulders as reaching far out sideways. Frequently during the day it is well to shrug them up and down and circle them forward and back to relax them, then let them fall softly into position reaching out sideways.

The Chest in Relation to Good Posture

If the shoulders are reaching out sideways the chest position is most probably correct also. In correct position, the chest extends out in front of the rest of the body—the sternum held high but not tensely so. Todd points out that the upper end of the sternum should be pulled up toward the head and there should be no forcing out of the ribs and diaphragm.[14] Never say to pupils, "Throw out your chest" for in all probability this will result in overdoing the command with an accompanying hollowing of the back. Say instead, "Sternum up and leading."

The Spine in Relation to Good Posture

The spine is the central interest of the skeleton as a whole. At its upper end (the cervical region) it supports the head; at its dorsal (chest) region it supports the twelve pairs of ribs which in turn support the sternum (breast bone) and clavicle (collar bone) which in its turn supports the scapulae (shoulder blades) and arms; and at its sacral (pelvic) region it is supported by the pelvic girdle which in turn is supported by the legs. In its normal condition the spine has three curves as viewed from front to back (anteroposterior view): The seven vertebrae of the cervical region are so joined together that they describe a curve forward; the twelve vertebrae of the dorsal region form a curve backward; the five lumbar (lower back) vertebrae repeat the forward curve and the five vertebrae of the sacrum are fused together and form a slight backward curve for the rear wall of the pelvic basin (Fig. 41).

The spine with its twenty-four joints has a great range of movement which is the combined action of many vertebral joints and not the isolated movement of one single vertebra on any one of its neighbors. The spine as a whole bends forward, backward, and from side to side or at any intermediate angle between these four directions, twisting and rotating as it bends. It can describe a great circle or turn on its

[14]Todd: op. cit., p. 165-166.

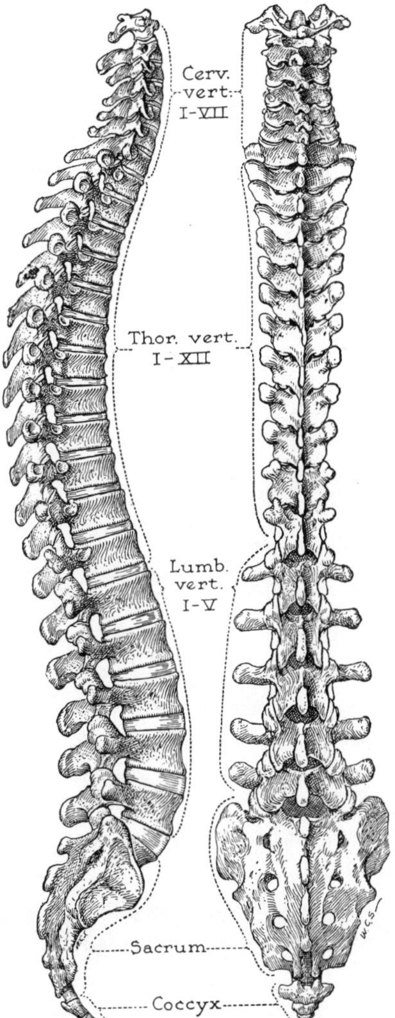

Fig. 41. The vertebral column as seen in lateral and posterior view. (Jones and Shepard: Atlas of Surgical Anatomy.)

own long axis. There are numerous variations of its possible movements.

Too frequently the anteroposterior curves of the spine are increased beyond the normal, resulting in the forward-head posture, the round back, and the hollow back. Muscle education should prevent or, if too late to prevent, should correct these increased curves before they cause health and postural damage.

From a rear view the spine normally is straight, but so many conditions are encountered by growing children which tend to throw the spine out of alignment that a large percentage of school children have some form of lateral curvature evidenced by one shoulder higher than the other, one hip more prominent than the other, the spine itself curving to the right or left in the dorsal region and most probably with a compensating curve to the opposite direction in the lumbar region, or with one total curve to the right or left involving both regions. The causes and treatment of lateral curvature are so complex that a person who has not specialized in the study of this condition should not attempt to diagnose or prescribe for it. Therefore postural training involving these difficulties will not be discussed in this book other than to point out that a strong, healthy body with strong musculature will be its own preventive for many such ills.

The Abdomen in Relation to Good Posture

The tilt of the pelvis has a definite and close relation to abdominal posture. It will be discussed later. As far as the abdominal muscles themselves are concerned they should be taut and not sagging. "Pull in your tummy" is a common and good admonition. The four layers of muscles that make up the abdominal wall form a natural corset, a wonderful one if only their owners would keep these muscles strong. These layers run horizontally, obliquely, and perpendicularly—the best of nylon girdles offer their three-way stretch in only one layer. There is nothing like strong abdominal walls for keeping a youthful figure.

The Hips and Pelvis in Relation to Good Posture

To many people the word "hips" refers to a vague, indefinite portion of the body. Many think of them as merely the upper edge of the ilium (large flaring-out hip bone) upon which you rest your hands when standing with "hands on hips." For our purposes we will consider the hips as that entire segment of the body below the waist at both sides and rear, rounding out over and including the buttocks. They are the side bulges, the shelf below the waist including the cushion nature gave us to sit on. In common parlance they are the "bustle," the "rear bumper," the "fanny."

The hip position is controlled by that of the pelvis. The pelvic cavity is a basin in the lower end of the trunk with its bony walls made up of the pelvic bones on the sides and front and the sacrum in the rear. The pelvic bones consist of three

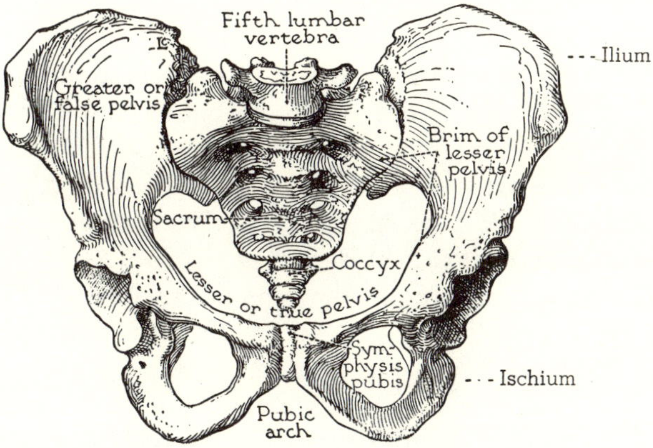

Fig. 42. Female pelvis, anterior view.
(Millard and King: Human Anatomy and Physiology.)

on each side, the pubic bone at the front, the large flaring-out ilium on the side, and the ischium tucked in under at the rear of the body. (See Fig. 42.) The lower end of the last named bone, the tuberosity of the ischium, is the part of the bony frame on which we sit. The sacrum is made up of five lower vertebrae of the spine fused together. The rear walls are normally higher than the front, but many people hold the pelvis in such a position that the floor of the basin is tipped so that the rear walls come up still higher and the front wall falls still lower. Figure 20 represents the pelvis tilted upward at the front in the correct position. Any further tilting of the pelvis downward at the front would result in poor posture.

When the pelvis is tilted too far downward the buttocks are thrust out prominently behind, the lower back is hollowed (sway back), the knees are apt to be locked, and the abdominal wall is stretched with the contents of the abdominal cavity seeming to hang out over a ledge, the ledge being the front rim of the pelvis, the pubic bones. In this position there is a definite downward pull on the abdominal walls, stretching and weakening them and tending to make them lose their elasticity. In this weakened condition fatty tissue readily accumulates.

In tilting the pelvis upward the buttocks are tucked in under, the lower back is somewhat flattened, the knees are relaxed, and the abdomen is drawn in to ride above the pelvic basin instead of hanging out over it. It takes much practice to acquire conscious control of the pelvic tilt, but since this is the hub of good posture and attractive carriage, it is worth any effort to acquire this skill. A few telling catch phrases make the task easier, such as "Tuck in your tail," "Hide your bustle," "Easy now, squeeze through this narrow space."

The pelvic girdle supporting the trunk, arms, and head must transfer the weight of this heavy superstructure down to the legs. It is admirably constructed for this task with the bones so formed and joined together that they make a bony arch with the sacrum as the keystone of the arch, absorbing shocks from body weights and transferring them through the sacroiliac joints to the ilia and thence to the hip joint and legs.

The Knees in Relation to Good Posture

In good posture the knees are relaxed and slightly bent so that they have "give" or "spring" to them. They are not thrust backward, tense, and "locked" as frequently occurs in bad posture. When the knees are "tight" or "locked" they react on the upper leg and hip so that the buttocks are forced outward and the whole body is tensed. A person who is inclined to stand habitually with locked knees should acquire the habit of speaking to himself many times a day saying, "Loosen up your knees" and frequently bending them imperceptibly and then trying to hold them thus "loose," ready for instant action.

The Feet in Relation to Good Posture

In good foot posture the toes point forward for better control[15] and the weight of the body is well distributed between the heel, the base of the great toe, and the base of the small toe with the main part of the load carried on the outer border.

The foot has three important functions, namely, to give support to the body, to absorb mechanical shocks and stresses produced by body movements and to serve as a lever for moving the body. The many joints of the foot permit considerable movement between the various bones, giving the foot a flexibility that permits it to absorb readily the shocks and jars resulting from body movements. Also its arches assure distribution of stresses and strains throughout the entire foot, thus easing the burden on any one part. Without this wonderful, shock-absorbing foot mechanism every body movement arising from the feet would transmit strain and stress to all parts of the body.[16]

The ability of the foot to give adequate support to the body when walking or standing whether it be at work or at play determines in an important measure how well the person can carry on his work or his play. The ability of the foot to do its work properly and without strain depends largely upon the tone and strength of the ligaments supporting the arches and of the muscles of the foot, ankle, and leg which support the body in a stable position above the foot. Therefore, the development and maintenance of strength of foot and leg muscles is a very important part of body conditioning and posture training. Special exercises to build and maintain proper foot

---

[15]If a growing child does not "toe straight ahead," it is far better for him to toe in than to toe out, since the toeing-in position is a strong-arch position and the toeing-out position almost invariably is a pronated position resulting in a poorly functioning foot. Beyond this the toeing-out position causes a turning out of the femur, throwing it out of proper position in relation to its junction with the pelvis thus producing strain there which is transferred to the sacro-iliac joint. "Pigeon toes" should not concern parents and teachers unduly. When the child acquires a desire to be attractive, he will take care of his pigeon toes, graduating easily into a "straight ahead" toe position, and in the meantime he has been protecting his longitudinal arch and sacro-iliac ligaments.

[16]For an excellent discussion of the normal functioning of the foot see Frank D. Dickson and Rex L. Diveley: Functional Disorders of the Foot. 2nd ed. Philadelphia, J. B. Lippincott Co., 1944, pp. 33-35.

condition and strength were described both in Chapters IV and V. A discussion of correct shoes to aid foot posture is given in Chapter VIII.

Although exercise for proper muscular development is important in foot hygiene, important roles in foot health are played also by proper nutrition, avoidance of excessive fatigue, and the wearing of correct shoes. The neglect of any combination of these precautionary measures may result in weak or pronated feet, with attendant pains in the feet, legs, and lower back and attendant outward and annoying symptoms of calluses, bunions, and corns.

Preliminaries to Teaching Posture Techniques

The teacher or leader who undertakes posture training should plan carefully the various steps to be taken in this work and their proper sequence in order to get the best results. For example, the stage must be set through giving pupils an appreciation of the values of good posture, thus instilling in them a desire to acquire good posture. With their interest now aroused, there are fundamental bits of information that should be imparted to them before practice starts, such as a brief explanation of the structure of the human body and the relation of the various parts to each other in terms of body alignment. The bare essentials for such information have been presented in the preceding material. When possible, the giving of the information should be accompanied by the showing of a skeleton or the use of large drawings or charts. Then come demonstrations and practice in the techniques of good posture followed by suggestions for out-of-class practice. The fundamentals of these teaching techniques are outlined in the material that follows.

Attitudes and Appreciations to be Developed

Why have good posture? The teacher will find material for a talk on this topic in the earlier part of this chapter under the various headings of spiritual, health, social, and economic values, and she will readily find much supplementary material to use in current physical education and health literature and from her own observations. She should give enough of this information to arouse the pupil's interest but not so much that she robs them of opportunities to get to work speedily at actual practice of good posture.

Information to Precede Practice

In addition to establishing correct attitudes and appreciations the teacher should give her pupils brief talks on (1) the skeletal structure of the body, (2) the function and action of ligaments and muscles which support the spine, shoulder girdle, chest, pelvis, legs, and feet, (3) correct alignment of body segments and (4) the basic principles of correct body mechanics. These talks should not be long. Sufficient material for this purpose has been given in earlier parts of this book. The talk on correct alignment will necessitate demonstrations of good posture by the teacher[17] or by pupils.

Visual aids through the use of films, slides, models, posters, and pictures of correct and incorrect posture are additional effective teaching tools, especially if the pupils' attention is called to various details of the models and pictures.

Practice Techniques for Attaining Good Posture

It is recommended that the teacher put her pupils through the following practices and in the order suggested.[18]

Standing Tall

1. "Stand in natural position. Close your eyes and try to get the muscle feel of your body in this position."

---

[17]It is to be hoped that the teacher herself has good posture. The good example of the teacher is the best type of visual aid for pupils. If she has been careless about her posture, intensive attention to it for a few weeks preceding the classwork on this subject should prepare her to face her pupils with confidence. Good posture is easily acquired if one has good musculature, but it is also easily lost with periods of carelessness.

[18]A warning is necessary here since standing still requires more effort than moving. To hold the body in a static position is readily disturbing as is well known from persons becoming dizzy and even fainting from standing at attention in parades, or when having clothes fitted, or when standing in a long line awaiting turn at ticket office and the like. Therefore, the teacher should not give too long a stretch of time to the practice of holding the standing position. Brief doses of such practice frequently repeated over a long period of time is better than large doses over a short period of time.

2. "Stand tall." Use no admonitions, merely say "Stand tall."[19]

3. "Relax."

4. "Now see that your feet are parallel and about 3 inches apart. Have your weight distributed as evenly as possible on the heel, base of great toe, and base of little toe. Check that the body segments support each other centered over the base of support. Now stand as tall as possible but without being tense. Shrug your shoulders and then let them settle down keeping them as wide as possible. Hold this position of being tall."

5. "Relax."

6. "Now with feet parallel and about 3 inches apart, keep the shoulders relaxed and stand as tall as possible. Hold it, close your eyes and try to get the muscle feel of this position through the entire body."

7. "Now try to get a feeling of 'upness' in the whole body—a feeling of lifting up all the body segments so they will not be so heavy on their supporting segments." (Since gravity tends to pull the body parts down, the correct body mechanics call for force opposing this downward pull, i.e., an upward pulling force, hence an important need in working for correct body alignment in the standing and sitting positions is not only keeping the body segments in vertical alignment with the pull of gravity, but also maintaining them in a feeling of "upness."

8. Repeat several times this alternate "relax" and "stand tall and up" with effort to get muscle feel.

Pelvic Tilt

To talk to pupils in terms of pelvic posture as such means sure defeat. There are interesting games to play which will teach proper pelvic control without throwing pupils into confusion in attempts to master anatomical terminology. Four good games are:

1. "Tuck in your tail." Have the pupils pretend they are

---

[19] By standing tall the curves of the spine are kept shallow and, although the curves of the spine opposing each other make for strength in the spine as a whole, that strength is weakened if the apex of the curves swing too far away from the line of gravity.

dogs with long tails and ask them first to make such a movement of the hips that they might thrust their imaginary tails out behind and then to move the hips in such a way that they might hide their tails between their legs. This game gives them an easy way to experience the feel of the pelvis in the two different positions and also increases their ability to control the tilt at will.

2. "Tuck in your hips." After the game of "Tucking in your tail" has been successfully mastered, then have the pupils try the same movements at the suggestion, "Tuck in your hips" instead of "Tuck in your tail." Work at this until they have learned this muscular control without trying to manipulate imaginary tails.

3. "Hold the coin." Have the pupils place one hand on the abdomen and one on the lower back and notice what happens to these parts when the pelvis is tilted downward, then upward, again downward, then upward. Then ask them to notice that in the downward tilt the buttocks seem to spread apart and then in the upward tilt are drawn tight together, so that in correct pelvic posture it is apparent that the buttocks should be firmly pressed together. The French army in its posture training of young officer candidates has them stand at attention holding a coin between the buttocks in order to acquire quickly this tight buttock position as an aid in controlling upward tilt of the pelvis. This can be turned into a game of holding an imaginary coin, loosening it, then grasping it, loosening it, then grasping it until the pupils have acquired this muscular skill and can at will tighten the buttocks.

4. "Squeeze through a tight space." Without mentioning pelvic posture, ask a pupil to squeeze through some tight space such as between two chairs placed close together and back to back and watch what she does; ask her also to take note. She will unconsciously turn sideways to push through, and if the space is still too tight for her ordinary standing position, she will unthinkingly tuck her hips under and pull in her abdomen to get them out of the way and will hold them tucked in until she is through the aperture. Incidentally, while she is thus "all tucked in" to be as slim as possible she is holding herself in her best posture. If she has habitual poor posture she should

spend much of her time squeezing through imaginary tight spots until she has gained easy muscular control of correct pelvic tilt.

Special exercises for stretching and strengthening tissues that prevent correct pelvic tilt were discussed in Chapter V. Since serious posture correction begins with pelvic tilt the pupil cannot spend too much time on practice at this skill.

Posture of Individual Body Segments

After pupils have learned to correct posture somewhat by standing tall and by pelvic control and have acquired definite muscle feel for corrected posture as compared to the muscle feel for their relaxed, natural, and probably poor posture, they are ready to be introduced to the specific posture of various body segments in order to ascertain wherein their particular weaknesses lie. To introduce this "breakdown" of posture too early is to confuse pupils. Have them first try for good posture as a whole, then as they sense specific difficulties they are ready to turn attention to specific parts of posture training, as follows:

1. Review carefully the material above under the heading "Good Posture" before proceeding with the practice.

2. Have the pupils practice correct position of various body segments singly, alternately relaxing to natural position and assuming correct position, trying to get the muscle feel of the correct one. Those pupils who have special difficulty should be given individual attention and, if necessary, they should be assigned special exercises for the part or parts that seem to be lacking in muscle strength or necessary coordination. (For these special exercises see Chapter V.) As each body segment is worked on, the teacher should give the pupil a brief explanation of the structure and function of each part and she should try to arouse the pupils' interest in and appreciation for correct posture of each part. Material for such talks has been covered in the foregoing pages.

3. Do not make this work too intensive nor too extensive in any one lesson. Expect to take a number of lessons to cover all of this material.

4. As the work progresses from part to part return occasionally to the practice of "standing tall," and "tucking the hips

under." This "whole" method of acquiring correct posture is of great value. The piecemeal method is used only to diagnose particular weaknesses in order to correct where correction is most needed.

5. For this segment-by-segment practice use the following admonitions:

HEAD
"Eyes front"
"Chin in"
"Chin up"

SHOULDERS
"Let the shoulders hang"
"Make them wide"

CHEST
"Sternum up and leading"

KNEES
"Loosen up your knees"

SPINE
"Keep your curves shallow"

HIPS
"Tuck your hips under"
"Hide your bustle"
"Spank yourself in"

ABDOMEN
"Pull in your tummy"
"Suck in your tummy"

FEET
"Arches up"
"Toes forward"

POSTURE AS A WHOLE
"Stand tall"
"Stand light as a feather"
"Keep a feeling of upness"
"Pull your body parts together"
"Let gravity work for you, not against you"
"Let your body hold you up instead of your holding your body up"

Body Alignment

Before starting classwork on this topic review carefully all the material previously discussed in this chapter. Then proceed as follows:

1. Vertical-line test. If possible to have at hand a supply of window poles, broom or mop handles, or rods of any kind that are as tall as the average pupil in the class, it would be well to check body alignment of each pupil by the vertical-line test. It is not wise to take class time to check each pupil one at a time.[20] If only one or two poles are available then demonstrate the pole test with one pupil, having the others note carefully how it is done so that they can try the test on each other outside of class and on their brothers and sisters at home. If there is a goodly number of poles available have the class work in groups and criticize each other's alignment. First, demonstrate as follows:

---

[20]Dr. Gertrude E. Moulton, at Oberlin College, successfully used as a substitute for poles a row of plumb lines suspended from a wire or tight cord, thus checking many pupils at a time.

1. Have the person to be tested stand side view to the class. Hold the pole at right angles to the floor with its lower end so placed at the side of the body that it is in line with the lower leg, passing just in front of the outer malleolus (ankle bone) cutting an imaginary line up through the thigh, pelvis, trunk, and head. It is not enough for this line to cut through the head just back of the ear to indicate good posture. One body segment may be out of line forward and compensated for by the corresponding backward swing of some other segment. All segments must be in line as indicated in the description of correct alignment earlier in this chapter. (Turn back to Fig. 38.)

It is well to give this vertical-line test in both natural and "stand tall, pelvis-tucked-in position" so that all may see how the various body segments swing into line as the correct posture is assumed. After a few repeated demonstrations on two or three different pupils the class members can test each other.

2. Wall practice. Have the pupils stand facing a wall with toes tight against it. Turn the head to one side and press the cheek and chest against the wall, then press the abdomen and knees also against the wall. Now keeping the chest, abdomen, knees, and toes touching draw the head backward, turning it to face front and hold it back so that the nose does not touch the wall. Now straighten the knees. As this action draws the abdomen away from the wall, place one hand on the abdomen and one on the hips at the rear. Now tilt the pelvis upward to swing the hips in under. Hold this position, let the arms fall to the sides and shrug the shoulders up and down a few times to relax them and then let them fall into easy position but keep them as wide as possible. Still holding the position, back away from the wall and walk about keeping as relaxed as possible yet holding the position of the head, chest, shoulders, abdomen, and hips, with knees relaxed. If pupils seem stiff have them "unloosen" and return to the wall and start all over again. If this is repeated often enough they will soon get "the feel" of correct posture.

Muscle Feel of Good Posture

Pupils should be told to close their eyes frequently as they strive for good posture and give careful attention to the sensations that arise within the body from various muscle tensions. With closed eyes the pupils should practice the following:

1. Standing, move head to right, then to left, then let it drop forward, and then draw it far back. Note the muscle feel of each position. Now open the eyes and place the head in the best possible posture, hold it, close the eyes, and concentrate on the sensations in the neck muscles with the head held in that position.

2. Repeat this process working on the arms and shoulders.

3. Repeat, working on one at a time: the chest, hips, spine, upper legs, lower legs, and feet.

4. Forgetting the various segments of the body, concentrate on the posture of the body as a whole. Stand tall, close the eyes and study intently the muscle sensations that arise from holding this position.

It would be helpful to do this work in couples, with one working while the other watches and admonishes until posture is correct, so that the worker is studying the muscle feel of the actual posture desired.

With practice a person soon comes to feel comfortable in correct posture and uncomfortable in a slumping one so that when one's mind is otherwise occupied the "muscle feel" alone will call attention to "getting off the track."

Standing with Weight on One Foot

When a person is tired from standing in one position the weight may be shifted but it should be done in such a way that the body segments support each other in correct alignment. Have pupils try shifting their weight in various ways to see for themselves which are the most conducive to continued good posture (Figs. 43 and 44.)

From the best weight-bearing, standing position with toes forward, feet side by side, parallel and about 3 inches apart, shift the weight as follows:

1. Hold foot position but shift weight to one foot. Invariably the knee of the free foot gives way and the hip sinks leaving the supporting leg to do the work of maintaining entire body balance with the hip on that side thrown far out of alignment to compensate for the low dragging hip on the other side. Still more fatigue piles up while in this posture.

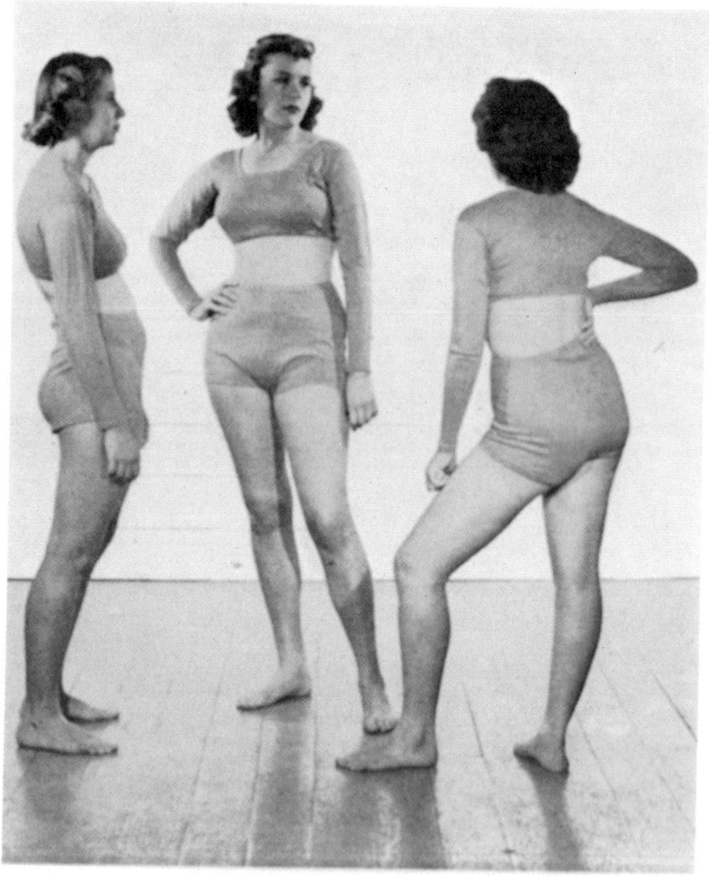

Fig. 43. Poor weight distribution. (Photograph by University of Nebraska Studio.)

2. Place the feet side by side but considerably more than 3 inches apart and with weight evenly divided between them. This is a somewhat tiring position and offers no particular advantages. In fact, it offers the disadvantage of the unattractive habit of keeping the feet apart. There are few things that make a girl look clumsy and awkward as readily as does a posture with the feet spread apart.

3. Place one foot a little forward but keep the weight evenly distributed on both feet. Good posture is still easy from this position.

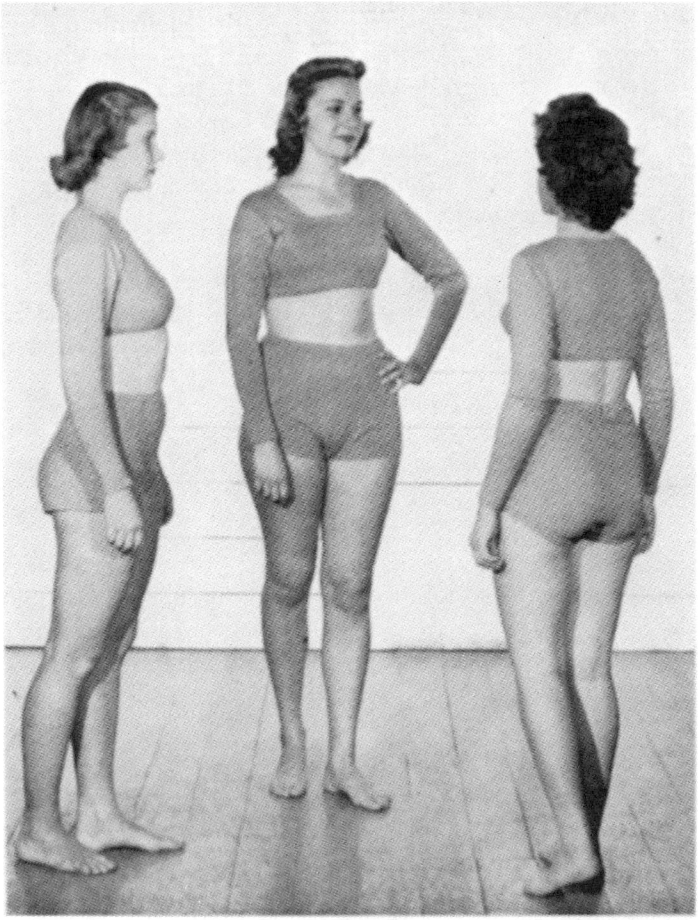

Fig. 44. Good weight distribution. (Photograph by University of Nebraska Studio.)

4. Keep this position but transfer the weight to the rear foot. Note that in this position there is a tendency to let the chest and shoulders slump. It takes real conscious effort to keep a high chest position with the body weight thrown backward. Such weight bearing is a "heavy" position. It is a posture of defeat, of "give-up." Actors invariably throw their body weight to the rear when they desire to give an impression of a person whc is lacking in force and energy.

5. Keep the position of Number 3 above but throw the weight on the forward foot. Note how easy it is to maintain an erect strong chest position. Actors realize the value of this posture in representing a person who is "on his toes," a person of vitality and energy. It is, as a standing position, a strong one if the weight is completely forward with only the toes of the rear foot touching the floor. This is a poise of readiness for action. It is a poise of confidence and hopefulness. It is easy in this position to be alert but not tense.

When standing in this position before an audience one can lean a bit forward on the rostrum and still keep the buttocks tucked in under, abdomen tight, and the chest and head up so that good carriage is maintained.

Since this weight-on-forward-foot position is so advantageous for good posture it is a safe rule to say, "Never stand with the weight on one foot unless it is in a one-foot-forward position and with the weight on the forward foot."

Posture and Heels

It is not sufficient to talk to girls about the effect of too high heels on posture: They should be shown what various heights of heels do to posture, and they should try the varying heights and have their attention called both to the resultant postural defects and their efforts at compensation. Blocks the varying heights of the heels which most girls wear on different occasions should be at hand, a large number of each of three sizes, (a) one-inch, the low heel of the walking shoes, (b) 2¼-inch, the Cuban heel, and (c) 2½- to 3-inch, the high heel of the dress shoe. Have the class work in twos, one checking her posture while the other, sitting on the floor at her side, slips the various blocks under her heels and criticizes her efforts. Have the pupils proceed as follows:

1. Remove the shoes and stand in the best possible posture. Hold this posture while the heels are elevated to take the one-inch block. Note any differences in muscle feel in ability to hold correct posture in the two situations.

2. Still holding correct posture repeat using the Cuban heel and note the differences.

3. Still holding good posture transfer to the high-heel blocks and note the muscle feel, also any shifts necessary in body

segments in order to maintain balance and correct posture. Note particularly that most of the pupils will feel the need of bending the knees to throw the weight of the body backward to compensate for the weight being thrown so far forward by the high heels. Note also feelings of tension in the lower back and leg muscles. Persons who have shortened Achilles tendons (the heel tendon) will find the higher heels comfortable. These people cannot go barefooted without a strain on the heel cord which in many cases causes positive pain. Each pupil should experiment and find out what height of heel makes it easiest for her to maintain good posture and then she should use that height on her walking and business shoes. A discussion of proper shoes is given in Chapter VIII.

Position of Feet in Standing in Relation to Good Appearance

The best way to have pupils learn for themselves what foot position gives the most attractive appearance is to have them demonstrate various positions before each other in groups. Figure 45 suggests the following:

Fig. 45. Common foot positions: good, bad and indifferent. (Photograph by University of Nebraska Studio.)

1. Stand with feet far apart sideways. Note that this gives a wide-hipped appearance, a stolid look.
2. Stand with feet tight together at both the heels and base of the big toes. Note that this gives a somewhat bow-legged appearance, even though it is a strong foot position.
3. Stand with heels together and toes turned out. The farther the toes are turned out the funnier the looks, the more the longitudinal arch falls, the more the foot is pronated. This is a weak foot position, and one that is a strain on leg muscles.
4. Stand with the heel of one foot at the instep of the other foot, toes slightly turned out. Note that this gives a small ankle

Fig. 46. The British Royal Family. Note the excellent posture of all five and the equally correct foot positions of Queen Elizabeth, Princess Elizabeth, and Princess Margaret Rose. (Permission of Press Association, Inc.)

appearance. This is the foot position used by women who care about their appearance when standing before an audience. (See Fig. 46.) As far as foot position is concerned it is by far the most pleasing in appearance. In this position the weight can easily be kept evenly distributed over both feet and it lends itself easily to correct body alignment. To rest in this position one needs only to shift the weight completely to the forward foot, letting the rear foot slide a little farther to the rear where it can relax momentarily while the posture of alertness is still maintained. Then the position of the feet can be reversed and the other foot and leg can relax for a while.

Information to be Given to Pupils Following Practice

It is important for pupils to understand that the few minutes of instruction devoted to posture training in any given week in classwork are but a mere drop in the bucket to the amount of time that they must attend to this work if habitual good posture is to be attained. They should realize that in the school class period they are taught the why's and how's of good posture, but that it is the practice that follows that makes perfect. To get sufficient practice, time must be taken out of other hours, since the physical education allotted time is all too short at best to do justice to any one of the many activities it must cover. Point out to each pupil that her body goes every place she does, it is with her every waking and every sleeping moment, and what she does with it during all the twenty-four hours of the day determines her habitual posture. Following are suggestions to give to her for living with good posture.

Everyday Activities

1. While standing at wash basin, dressing table, work table, ironing board, or anything else that necessitates bending over at a task, keep one foot advanced with the weight on this forward foot so that there will be a broader base to give greater stability with less strain in maintaining body balance. On the other hand, when the feet are kept together it is the path of least resistance to thrust out the buttocks as the trunk bends forward in order to preserve balance: Accompanying this are relaxed abdomen, locked knees, hunched-up shoulders, and neck thrust forward.

2. Standing with arms akimbo hunches the shoulders up and forward and spoils the neck and upper back lines. Standing with even one hand on hip with the body twisted as is usual in such an attitude, there is hunching of the shoulder on that side.

3. When being helped into a coat it is important to keep erect and not squirm and twist. The arms alone can do the work. One foot should be advanced with the weight on the forward foot to keep good balance.

4. When standing for a long time in one place, if the situation permits, use a rubber or cork mat or folded rug for a cushion under the feet. Also shift the position frequently with the right foot forward taking the weight, then the left forward taking the weight, then the feet together with both taking the weight.

5. Standing on the streets waiting for a street car or for light signals to change or waiting in long lines offer excellent opportunities for special thought to posture. All too frequently such moments are mental blanks yet they may be turned into golden moments to practice good posture.

6. Other valuable but usually lost moments for posture work are those spent on "hair-do's" and primping before a mirror. That mirror should show other things than just hair-dress and make-up.

Constant Watchfulness

Use every possible means to make the pupils realize that if they are to attain and maintain good posture they must give conscious attention to it at home, at school, at church, on the streets, on the playfield, at shows, while working, and while playing. Dozens of times a day pupils should check their posture and correct it as necessary and if they have difficulty with any particular part perhaps they should take special exercises for certain muscle groups which need strengthening. They should be made to realize that once body musculature is in good condition the best way to attain habitual good posture is to keep oneself in habitual attitudes of good posture. Nothing can train a person to "stand tall" so readily as standing tall. This is just plain, everyday horse sense.

Will Power and Determination

Many people fail to have good posture not because they have weak muscles but because they have weak wills. Good posture is largely a frame of mind, a set of the will. Any person of average muscular development can readily have good carriage if she has determination and will. If she determinedly thinks of herself as a confident, vital, energetic person, she is apt soon to take on the attitudes of such a person. The more she is able to assume that attitude the easier it is to convince herself that she does have those qualities, and the more she actually acquires those qualities, the easier it is for her to express them through her posture. Thinking good posture and acting good posture pursue each other in an ever widening circle.

Teaching Suggestions

Once you have taught pupils the fundamentals of good posture start every activity (sports, dancing, swimming, gymnastics, marching) with the admonitions "Stand tall," "Tuck in your hips," "Pull in the abdomen" and repeat these frequently throughout every class hour. As you pass pupils in corridors or on the street notice their carriage and, if they are slumping, say to them quietly, "Stand tall," or better still, have them watch each other and when necessary pass on the admonition.[21]

WALKING

Walking involves first the overcoming of inertia of the standing position and second a constant propelling of the body weight forward from one base of support to another with the ever present need of maintaining stability of the body by keeping the body segments centered in line with the pull of gravity over the ever changing base of support. This base is very small compared to the mass which must be supported and balanced above it. This offers difficulties. It is most interesting to sit in a metropolitan railroad station or stand on a busy street corner and study the gait of the people passing by. The few look effective while the rest, the great majority, look inept as they propel themselves in a great variety of peculiar ways. Some waddle along, others lurch past, some slink on their way, and others seem to breeze by. Most drag along after themselves reluctant shoulders.

The body build, which a person inherits and about which he can do little, has certain decisive influences upon the manner of his walking, but beyond this there is much that he can do to control his gait.

Some physiologists speak of walking as a constant falling forward, but Mensendieck describes the correct walk as a pulling forward with first one foot, then the other, drawing the

[21]At the University of Southern California the Department of Physical Education for Women has parodied a muchly aired cigarette slogan into S T — T Y H meaning "Stand tall, tuck in your hips" and the college girls call out this secret watchword to each other as they meet in classroom and on the campus and catch each other slumping.

body onward by a "suction" step.[22] She contrasts this "suction" or pulling step with the commonly used "propulsive" or pushing step in which the weight is pushed from the rear foot to the forward one. In the "suction" step the weight is kept to the front and the step is light. With the chest seeming to be pulling one along this type of walking gives a floating, gliding effect. It is the gait so many leading actresses employ. In the "propulsive" step the weight is back on the rear foot and the gait seems heavy.

Gait and Posture

Posture has much to do with gait. A good posture means the battle half won for a good gait. Hence it is important in walking to maintain an excellent posture.

Walking and Conservation of Energy

Any type of walking that uses as little movement as is absolutely necessary and that lifts the body as little as possible is the most efficient type since it uses up the least energy. Increased arm swing and body lifting add nothing to the main object of propelling the body forward in walking, even though they do have their purpose in running, skipping, and like activities.

The Mechanics of Walking

The teacher should explain and discuss with the pupils those principles underlying the effective use of the body which are related to walking and from these basic principles should lead them to an understanding of the general rules of walking such as:

1. Good carriage should be maintained. (If the pelvis is kept in a good upward tilt position in the front, walking with a jarring effect is almost impossible.)
2. There should be a free swing of the legs from the hips.
3. There should be a strong pull of the weight onto the forward foot.
4. The toes should point straight ahead.

---

[22]Bess Mensendieck: It's Up to You. New York, Mensendieck System, 1931, pp. 186-187.

5. There should be only that arm swing which comes from body momentum. (When time is to be conserved and the walking is to be at a fast tempo, the arm swing should be increased, thus serving as an aid in the forward thrust of the body.)

After this the teacher should take up specific problems of walking, such as are listed in the material that follows.

How Do You Walk?

1. Do you pound the floor when you walk or do you float along?
2. Do you toe out or forward?
3. Do you drag your hips along protruding far out behind you or do you keep them tucked under?
4. Do you lead with your hip bones, leaning backward with your shoulders dragging along behind you or do you keep your shoulders over your hips?
5. Do you lean forward from the hips and "just go to it," pushing the sidewalk away from you to the rear? It is easy to stumble and fall forward from this gait as the body is already pitched forward.
6. Do you walk with the feet too far apart? The knees should just miss rubbing each other unless you were born with a particularly wide pelvis and the thigh joints are farther apart than in most people.
7. Do you walk with stiff knees or do you keep the knees relaxed?
8. Do you mince along, with too short steps, stiff and tense, afraid to step out, or are you a strider, taking too long steps and swaying the shoulders and arms?
9. Do you walk tensely, tightly, nervously or do you keep relaxed?
10. Do you attempt to walk gracefully in heels that are too high for your build?
11. Do you swing your arms vigorously or just let them move with the body motion?

The Graceful Versus The Ungraceful Walk

A few years ago a group of nine hundred university women voting as a lay student-judging committee made a study of

Fig. 47. How do **YOU** walk? (Photograph by University of Nebraska Studio.)

graceful and ungraceful walk.[23] They decided that the ungraceful walker tends as follows:

1. To carry her arms back of her body.
2. To carry her arms in a jerky fashion.

[23]Arranged from Ruth Bass: A Study of Graceful Walking. Research Quarterly of American Association for Health, Physical Education and Recreation, May, 1937, pp. 177-78.

3. To have a peculiar lower arm carriage.
4. To have an exaggerated position of the head.
5. To zigzag parts of the body.
6. To exaggerate movements sideward and up and down.
7. To have stiff feet.
8. To either hold the trunk rigid or hold it forward.

They agreed that the graceful girl tends:

1. To move the various parts of her body in harmony.
2. To move so that the tip of the shoulders moves in a wavy line rather than in a zigzag manner.
3. To carry her arms in line with her body, swinging them as much in front as in back, and having them fairly in line with the body when the feet pass each other.
4. To have a vigor of arm and leg swing.
5. To swing her legs from the hips.
6. To lift the chest and head easily and strongly, giving an impression of "upness."
7. To place the heels in a direct line, one with the other, and to transfer the weight forcefully, but easily and smoothly, from heel to toe.
8. To carry the head, shoulders, hips, knees, and ankles in good alignment.

In this study it was found that some were ungraceful in certain aspects of their gait, others in other aspects.

Practice Techniques for Correct Walking

To teach correct walking the instructor needs to develop in the pupils an appreciation for correct walking, to ascertain the correct length of stride for each pupil, to analyze for them the fundamentals of correct walking and to criticize constructively the natural walk of each pupil and her efforts at improvement.

Developing Appreciations

The foregoing information on correct walking should be given to the pupils in a preliminary talk, followed by demonstrations of correct and incorrect walking. A child with a poor bearing should not be singled out to display her poor skills: instead, call for volunteers to demonstrate various types of walks with the class criticizing each in terms of basic principles.

Length of Stride

Before class the floor should be chalked off with lines across half the width of the room and one foot apart for a stretch of at least 12 feet, with the spaces between numbered 1 to 12. During the class hour the following procedure should be carried out:

1. Have one pupil walk in his natural gait over the spaces while the teacher notes the spaces his steps fall in and reports it to the class, interpreting it into terms of stride length. Repeat this with one or two other pupils who are either taller or shorter than the first one so that the class will observe the reading for a variety of lengths of strides and heights of persons.

2. Have the class divide into couples.

3. A's line up on the starting line and take a trial walk for their observers. Then after all the questions that arise have been answered, have them walk for an actual measuring of their stride.

4. B's report findings to A's.

5. B's have their strides measured and announced to them.

6. Check to find out the length of the shortest and the longest strides recorded.

7. Have all whose stride measures the shortest figure line up and walk down the space together, noting whether any are noticeably taller or shorter than the most. Make note of these.

8. Repeat Number 7, calling for those whose stride fell within the next grouping, and so on, through various groups to the longest stride group. Divide these groups roughly into only three sets since this will suffice to weed out the "off-stride" pupils. Using more sets will be too time consuming.

9. The list of "off-stride" pupils should be saved for later reference should it develop that these persons have other walking difficulties. A stride that is too long or too short for one's height might easily cause awkward gait.

10. If on later observations these "off-stride" pupils need to alter their stride for more effective walking, they should do individual work, practicing assiduously at walking in spaces blocked off for the length of stride used by the pupils of their own height and same relative leg length who are judged to be good walkers.

Slow Motion Walk

In order to acquire an understanding of the component parts of walking the teacher should demonstrate walking in slow motion and then have the pupils imitate and practice it as follows:

1. A step divided into its component parts.
   a. Stand with the body segments in correct alignment.
   b. Take one step forward and hold with the weight evenly distributed between the two feet.

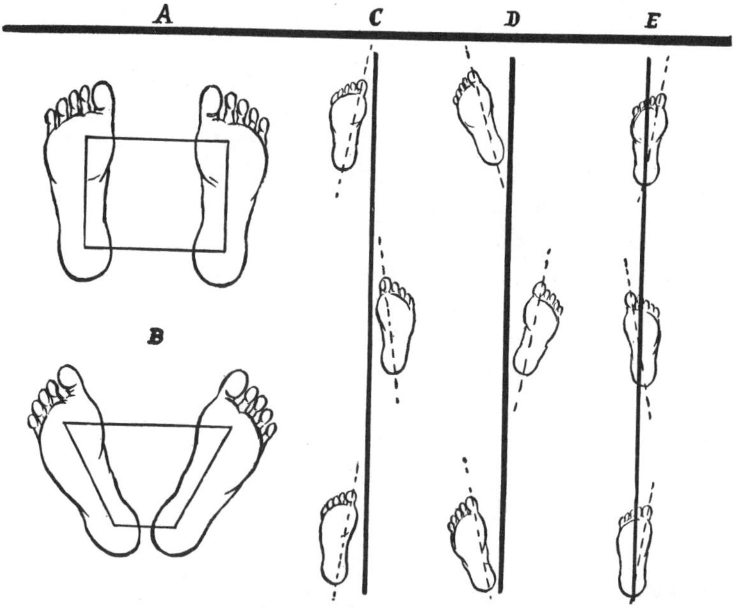

Fig. 48. A, Four square position, standing, which is correct. B, Military position, standing, incorrect. C, Four square position, walking, correct. D, Military position, walking, incorrect. E, Indian position, walking. (Courtesy of Thomas S. Ellis.)

   c. Check that the toes point straight ahead and the feet are only as far apart from side to side as is necessary because of the conformation of the body, and only as far apart forward and back as an average length of step requires (Fig. 48).
   d. Transfer the weight to the forward foot with the heel of the rear foot off the floor and only the toes touching.

e. Hold the weight on the forward foot as you relax the rear leg, bending the knee to let the thigh swing forward. As this movement proceeds, relax the foot and ankle and keep the legs so close together that the knees just graze each other.

f. Still holding the weight on the forward foot, lift the thigh enough so that the relaxed foot, that is hanging downward from the relaxed ankle and the bent knee, can clear the floor, then straighten the knee, thus swinging that foot forward in front of the foot holding the weight.

g. Flex the ankle of the raised foot, swinging the weight forward from the arch of the supporting foot to the toes and finally transfer it onto the heel or arch (depending on the length of the stride) of the foot that has just been swung forward, gradually letting the entire foot take the weight as the heel of the rear foot gradually comes up off the floor.

2. Practice this in slow motion to rhythmical counts.

3. Practice in increasing rhythm until the class is walking in its ordinary walking rhythm.

As practice moves along the teacher should admonish the pupils for correct movement, and she should halt the class frequently and return to the original slow motion practice.

Warning To Pupils

1. Be careful not to sway.

2. Point the toes forward.

3. Place each foot straight ahead of its last imprint. In other words, do not walk on one imaginary straight line but on two, one for each foot, with the two lines only as far apart as the distance between the center of the two thighs.

4. Be sure to relax the knee and ankle and let the foot hang limp when being lifted.

5. Keep a correct posture.

6. Step lightly. Be buoyant—hold the weight of the body high.

7. Let the thigh lead, not the abdomen.

8. Don't drag the legs after you.

Practice Drills

Have the pupils practice as follows:

1. Practice both the "push-off" and the "pulling" step until you sense the "muscle feel" for each. Then try to make the "pulling" step the habitual step.

2. Practice walking on straight lines. (Ziegfeld's training for his show girls.)

3. Practice walking with a block or book on top of the head. (The supposed-to-be time-worn techniques used in the training of every princess.)

4. Practice walking as in a grand march. Use a good marching song. Keep the attention centered on the erect carriage and correct techniques of walking.

5. Stand against a wall in good posture with the feet a short way from the wall and the back touching it. Then, without shifting any body parts, walk away from the wall holding the good posture.

6. Occasionally return to the slow motion walk and check on the details of the movement and make corrections as necessary.

7. Strive to get the "muscle feel" of correct walking through the entire body.

8. Have the class march a little at every class period so that walking postures can be checked. Inspect every pupil critically as she marches past you. Such practices are a waste of pupil's time if they are not to receive constructive criticism constantly.

Walking Tempo

Both low and high speeds of muscle contraction are wasteful of energy, therefore fast walking or the opposite, sauntering, lower efficiency of movement. The teacher should search out those who for their natural gait walk in these two extremes of speed and should give them special drill to help them to tone down or tone up their walking tempo, whichever is necessary.

Practice Outside of Class

Urge the pupils to give constant attention to correct walking. Every time a person walks any place—whether it be merely across a room, to a class, or on a long errand, that is the time to practice. It doesn't require special time. The best way to learn how to walk correctly is to walk correctly every time one walks. In addition to this the following suggestions to pupils may be helpful:

1. Analyze the walking manner of people whose gait you admire. Much may be learned by studying the movements of famous actresses. They have been drilled for months and years in correct body mechanics, and their success depends largely upon this skill. Perhaps such skill would materially increase your success in life.

2. Practice correct walking in various heights of heels. Note which ones throw the body seriously out of balance, preventing good carriage, and then avoid those heights of heels for walking.

Walking For Long Distances

Hiking is one of the most healthful of physical activities and it is easily a most enjoyable one provided one knows how to walk correctly for long distances and how to dress appropriately for the exercise. To prepare the students for such activities the teacher should present the following material:

Techniques for Long Hikes

1. Maintain good posture and thus ward off too early fatigue.

2. Take a slow gait but one with vigor to it. Four miles per hour is a good rate for level going, still slower if on rising ground.

3. Every ten minutes rest one minute on your feet.

4. Every hour rest ten minutes seated or reclining.

5. Every four or five hours rest one hour, lying down 15 minutes before eating and relaxing the remainder of the hour after eating, preferably flat on your back with eyes closed.

6. On highways walk on the side of approaching traffic.

7. Shoes for long hikes should be one size longer and wider than street shoes to accommodate woolen hose and to allow room for swelling of the feet under continuous weight bearing.

8. To build up ability to walk long distances easily or to walk a great deal practice the following:

a. Every day for a week walk a half-mile (6 blocks).
b. The next week walk one mile (12 blocks) two or three times.
c. The next week increase the amount to 1½ miles (18 blocks). Do twice that week.
d. The next week do 2 miles (24 blocks).
e. The next week do 3 miles (36 blocks) and keep it up weekly.

9. Once a week take a 3- to 5-mile walk.
10. Once a month take and 8- to 10-mile walk.

Motivating for Long Hikes

1. Keep class lists posted with space after each name to record the length of hikes taken each week.

2. Keep records of the longest single hikes and also records of the longest accumulated distances over given lengths of time.

3. Have contests for cumulative hikes to imaginary interesting places. If a certain city is 50 miles away, have the class keep record of their hikes of 5 miles or over and see who first reports on an accumulated 50 miles and thereby is declared "the first into Shangri-La."

CLIMBING

Stair climbing when done properly is excellent exercise, also, an excellent hip reducer since the hip muscles must lift the entire body weight almost straight upward against the pull of gravity. If sufficient stair climbing is indulged in with correct techniques, the hip muscles soon lose flabbiness and become firm. To maintain this firmness one has only to continue much correct stair climbing.

It is not only an excellent leg strengthener but it is also an excellent organic stimulator. It takes 15 to 17.7 times as much energy to climb an ordinary flight of stairs as it does to walk on the level a distance equal to the vertical height of the flight of stairs,[24] hence the quick feeling of muscle fatigue in the leg extensors and the quick breathlessness when climb-

---

[24]Edward C. Schneider and Peter V. Karpovich: Physiology of Muscular Activity. 3rd ed. Philadelphia. W. B. Saunders Co., 1948, pp. 59-60.

ing. To descend stairs takes one-third the energy of ascending[24] but even that is more than is required for walking on the level.

To be able to climb stairs without puffing and panting means good cardiorespiratory endurance. When climbing makes one puff it should be done more slowly, giving the heart action more time between steps to catch up with the increased demands made upon it.

The Effective Use of the Body in Climbing

When climbing is done incorrectly the weight is lifted without the full assistance of the muscles that are best adapted tc do this work and results in undue effort; therefore it is important to use the body correctly in climbing. This means adherence to the underlying principles of those body mechanics which relate to lifting and lowering the body. Translating these principles into terms of climbing, we arrive at the following basic tenet: The body segments should be kept as directly above the base of support as is possible, yet allowing for the thrust of the body in the desired direction.

After this tenet is thoroughly discussed, the teacher should then break down the subject into class instructions such as are presented in the material that follows.

Mechanics of Ascending and Descending Stairs

1. Hold the body erect. Stand tall, pull in the abdomen and tuck the hips under (Fig. 49).

2. Incline the entire body slightly forward to assist in the shifting of the weight since the body is to be pulled forward as well as upward. The forward inclination should include the entire body from ankles up, not just the trunk.

3. Plant the whole foot (not just the ball of the foot and toes) firmly on the tread of the step above (not just on the edge of the step).

4. Straighten the bent knee and thigh of the uppermost leg. This will force the body up to a position over the forward foot, leaving the other leg free to bend the thigh and knee so as to swing the foot up onto the next step.

5. If there is a hand rail, use it for whatever assistance it may be as a balancer, but as a balancer only. Make the legs do the work.

Fig. 49. Stair climbing. (Reproduced from WAC Field Manual by permission of the Department of the Army.)

6. Ascend neither too slowly nor too fast. Both rates waste energy. Experiments prove that the rate of 1.3 second per standard step is the one demanding the least energy.[25]

7. In descending stairs keep the body erect, have the free foot forward to hang out over the next step, bend the knee and ankle of the supporting leg, shift the weight slightly forward and, keeping it under control, let it down onto the forward

[25]Schneider and Karpovich: op. cit., p. 60.

foot as it contacts the step below with the foot placed squarely in the center of the tread. Be careful not to drop the weight suddenly onto the lower foot. The movement should give a flowing rather than a bouncing effect.

Mechanics of Climbing Ladders

1. See that the ladder is placed firmly in a slanting position against the supporting object and is well balanced.
2. Face the ladder, standing erect with hips tucked under.
3. Grasp the ladder well overhead.
4. When stepping up with the right foot have the left arm uppermost; with the left foot, the right arm uppermost.
5. Look up at the rungs of the ladder.
6. Keep the body out away from the ladder.
7. Keep relaxed.
8. Place the foot on the rung, touching back of the ball of the foot, not on the ball or the toes.
9. Have the foot firmly planted on the rung before releasing the grasp.
10. Slide the opposite arm up to a higher grasp as the leg extensors push the body upward.
11. Push up with the legs; don't pull up with the arms.
12. Use both hands. Do not carry tools or equipment by hand unless absolutely necessary. Carry them in the pockets in your clothing or in a carrier thrown over your arm or have them handed to you after you are up.
13. Take your time. Don't rush.

Unsafe Practices in the Home

Since so many accidents to women occur in the home from falling from chairs, stools, and ladders, it is evident that there has not been sufficient training in correct use of the body and in safety measures for climbing. The teacher should warn pupils against the use of chairs, chair arms particularly, and stools as a substitute for ladders. When a chair or some other piece of furniture must be used in lieu of a ladder it is important that the chair be completely stable and that the foot is placed firmly in the center of the seat before any part of the weight is transferred to it. Most accidents occur from attempts to transfer the weight of the body to a foot precariously placed on the edge of a chair or on a chair arm.

Practice Drills at Correct Climbing

For stair climbing, take the class to a flight of stairs; for ladder climbing, procure various types of ladders frequently used in homes and drill the class in their actual use.

1. Explain and demonstrate correct climbing.
2. Have the class practice in couples.
   a. Climbing stairs.
   b. Climbing ladders.
3. Warn against unsafe practices in the home.

Suggestions for Out-of-Class Practice

Urge the pupils to turn every need to climb stairs or ladders that arises throughout the day into an opportunity to practice correct climbing.

RUNNING

To watch the average group of grown women attempt to run is evidence of the neglect of their physical education when young. Many of life's situations would be easier for grown-ups if they possessed skills for easy, effective running. Children run a great deal of their own volition, but far too many of them do it ineffectively. A little time spent in school on the techniques of correct running will mean ease and comfort instead of stress and strain both for the child's play periods and for later life situations when quick dashes must be made for street cars and buses or to get out of the way of hazards.

The Effective Use of the Body in Running

The main differences between walking and running are that in running there is a longer stride, a greater swing of the arms, a "take-off" from the toes instead of the whole foot as in walking, a little more knee action, and a little more inclination of the trunk forward. At each running step both feet are momentarily off the floor. For good running, good posture is paramount. The techniques of teaching correct running have been discussed in Chapter VI.

For the conservation of energy a person should not run if

walking will get the job done quickly enough. In fact, one never should speed at any work if a slower rate will be satisfactory.

JUMPING

For most people the necessity for jumping arises but seldom in everyday living although there do come the unexpected moments when a ditch or a mud puddle must be cleared or when it is necessary to jump from a platform or other high object. Since most boys, sometime during their school days, go in for jumping as an activity for its own sake and are interested enough in skillful performance to practice correct techniques, they are usually sufficiently skilled for whatever emergencies may arise requiring jumping. But girls on the whole have not had such an urge in their growing years, so that, unskilled in these techniques, they are handicapped when such emergencies arise. Therefore, the physical education department should teach this technique to all girls. (Jumping as considered here is not to be confused with those forms of jumping that are a part of track events in sports activities. The fundamentals of broad and high jumping belong to a course in sports fundamentals and are not taken up here.)

The Effective Use of The Body in Jumping

Practice at jumping should involve at first merely jumping forward on a level place with attention focused on landing techniques. Jumping from heights should be attempted only at gradual distances and only after correct techniques are mastered. For classwork, heights beyond those of an ordinary stool or bench need not be attempted. There is doubt on the part of some educators about the wisdom of girls jumping from heights because of the supposed inherent weakness of the pelvic floor.[26] Most certainly they should not practice at such activities during the menstrual period. It is sufficient that they be drilled in the art of correct landing from a low jump, leaving the real emergency to test the skill. And it will test safely if the girl has been carefully drilled and frequently reviewed in the techniques of jumping from a low height.

[26]Research tends to indicate that displacements of the uterus rarely occur unless the pelvic floor is weak from childbirth. (Dawson: op. cit., p. 494.)

The students should be instructed in those underlying principles of body mechanics which apply to jumping.

Mechanics of Jumping

With the following instructions, the classwork is now ready to be started:

1. Maintain good posture. Stand tall, pull in the abdomen, and tuck the hips under.
2. Take a crouch position (the greater the jump desired, the deeper the crouch). To do this, rise on the toes, bend the knees and hips, incline the trunk forward, and swing the arms backward.
3. Vigorously push off with the toes, extending hip, knee, and ankle joints and swinging the arms vigorously forward to help propel the body forward.
4. Land with "soft" knees and ankles so that there is much give in both sets of joints as the body lands. This is very important if jarring of the whole body is to be avoided. Landing without this "give" in the knees and ankles may readily cause sacro-iliac strains, particularly in women.

Practice Drills

After the fundamentals have been explained the teacher should put the class through drills as follows:

1. Jump in place, getting the "muscle feel" of soft knees and ankles.
2. Jump in place, alternately landing with feet apart and together, concentrating on proper landing.
3. Jump forward, concentrating on the crouch and push-off but not neglecting the correct landing.
4. Run a few steps and then jump, taking off from the right foot.
5. Repeat Number 4 from a left-foot take-off.
6. Jump from a bench, lower steps of a stairway, or a ladder. Be sure that the movable objects are held firmly in place.

SITTING

Practically everyone spends a great deal of his waking hours sitting, and for the proper functioning of the body this sitting should be done correctly. The next time you go to a movie where one of the better actresses is being shown, study carefully her sitting posture and poses. Also note how she approaches a chair, how she lowers herself into it, and how she rises from it.

The Effective Use of the Body in Sitting

Maintaining good position is of course the keystone to correct sitting, but the use of the body with the weight off the feet poses a few problems not involved in standing. However, the basic principle of keeping the body segments in vertical alignment over the base of support applies here as well as in standing, even though the base of support has been transferred to a location nearer the center of gravity of the body. After a review of this basic principle, the teacher is ready to lead the class into a breakdown of the general principle into the specifics of correct sitting. For this she should instruct them as follows:

The Mechanics of Correct Sitting

1. Keep an erect posture of the trunk and head. Sit tall, pull in the abdomen.
2. Sit on the tuberosities of the ischia (the lowest of the three bones on each side that make up the pelvic girdle). This prominence can be felt readily on each side when you sit down on your hands. This position will hold the trunk directly above the pelvis. If, instead, you sit on the lower part of the spine the buttocks are spread out tending to make the hips broad.
3. Sit far back in the chair.
4. As soon as you are settled in the chair give a quick thought to the "muscle feel" of your posture and correct it as necessary. If you start out in a good position you are apt to retain it for a fairly long period of time.
5. If remaining seated for long, check your posture occasionally to remedy any slumping that tends to overtake you.
6. Keep the posture relaxed, not tense, and thus avoid fatigue which, in an effort to get away from the tensions, causes slumping.
7. Keep the feet and knees together whether the feet are firmly planted in center front of the chair or to one side, either somewhat forward or backward.

8. Whether the feet are placed in front or to one side, keep the knees as close together as the conformation of the thighs permit. In other words, don't sit sprawling in a chair with the knees spread apart.

9. If, in sitting for long periods, you wish to change the position, it is mechanically correct to cross the legs provided you do it with both thighs taking turns at supporting the weight.

10. When sitting with the lower legs crossed, cross them as far down as possible and without pressing the supporting foot tensely into the floor. Do not cross them with the feet extended out in front unless you are sure they will not in that position cause inconvenience to others. Usually they are less in the way of others if, when crossed, they are swung to one side. Keep the ankles relaxed. (Review Fig. 46.)

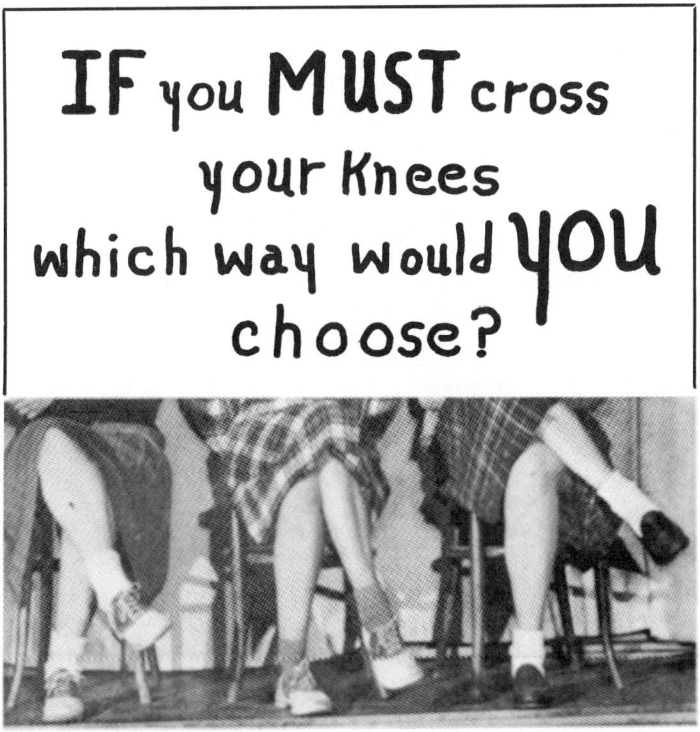

Fig. 50. Suggested poster for class motivation. (Photograph by University of Nebraska Studio.)

11. If you cross the legs cross them either above or below the knee, not at the knee (Fig. 50), for this position causes the uppermost foot to swing far out in front in the way of all who come near and causes the flesh of the calf of the uppermost leg to bulge in an unsightly manner. When the legs are crossed above the knee the upper leg can hang down easily relaxed and almost parallel with the other leg, hence inconspicuous and free of bulges. A generation ago it was considered "unladylike" for a woman to cross her legs, but if done either above or below the knee and while maintaining good posture and not drawing the skirts up too high, it is acceptable today. This position should never be assumed in public if the skirt is very short or narrow. (If women had more full-length mirrors at hand they would be amazed to see how unattractive they usually look when sitting. This is chiefly because of knee and foot positions.)

12. Keep the hands relaxed and in the lap or on the chair arm unless they are occupied with some work. Do not sit with one or both hands on hips for this tends to hunch the shoulders.

13. When necessary to reach or bend forward while seated, rock forward from the hip joint, thinking of the tuberosities of the ischia as rockers. Don't curve the spine forward. Keep the trunk and head in straight alignment with the forward movement originating in the hip joint.

14. While at work in a sitting position have the hips against the back of the seat (Fig. 51). If when in this position you cannot reach your work without much bending forward you need to sit nearer your work. To do this place a support at your hips, a large pillow or back rest, and then keep your hips against that support.

15. Let the shoulders hang wide and relaxed. If it is impossible to do so while seated and at work, the seat needs some adjustment. You should work with your elbows close to waist level; anything above that hunches the shoulders.

16. If, in order to reach your work the seat has to be adjusted to a height that lifts the feet from the floor, you should use a foot stool so that the feet may be firmly planted on a support.

Fig. 51. (Courtesy of the Department of the Army.)

The Mechanics of Lowering Oneself Into a Seat of Ordinary Height

1. Approach the chair and as you reach it turn your back to it, being careful that you are standing close enough so that the calf of one leg touches it while the other foot is a little bit advanced and the body weight thrown onto the forward foot.

2. Keep an erect posture. Stand tall, pull in the abdomen, tuck the hips under.

3. Bend the knees and hips, inclining the trunk slightly forward and letting the body down onto the seat; transfer the weight gradually onto the rear foot.

4. Do not bend forward and seem to double up as the trunk descends, giving the effect of groping for the seat with the buttocks. Also do not grope for the seat with the hands. If one leg is touching the chair you know the chair is there.

5. Do not drop into the chair. Keep the weight under control. The leg extensors should stay on the job and permit the hips to flex gradually and smoothly, thus letting the trunk down gradually and smoothly.

6. If you stand close to the chair with one leg touching it you can let the body down into the whole seat with the hips well back against the back of the chair. Don't be seated on the edge and then push back into it. At the start be seated where you are going to sit.

The Mechanics of Rising From a Seat of Ordinary Height

1. Place one foot a little to the rear with heel raised preparatory to push off and receive the body weight, the lower the seat the farther back the foot.

2. Keeping the head and trunk in correct alignment, incline the trunk a little forward, push off from the toes of the rear foot and extend the hip and knee joints, thus lifting the body to an upright position.

3. Straighten up to a good standing position transferring the weight to the forward foot thus moving the body away from the chair and leaving the rear foot free to step away from the chair.

4. Do not help yourself out of the chair by using the hands and arms. Let the leg extensors do the work. Sometimes seats are so low that one must have assistance before the leg extensors can "take hold" but as a rule people "help themselves" out of chairs unnecessarily. Often people put the hands on the thighs to help push the body up by the arms. Although this may be of assistance to the back muscles and thigh extensors it is unattractive.

Practice Drills

Have stools, benches, or chairs enough for half of the class and have the pupils work in couples, one to practice while the other criticizes as follows:

1. Approach the chair in good posture.
2. Turn at the chair and be seated.

3. Remain seated to receive criticism.
4. Arise from the chair.
5. Repeat Numbers 1 to 4 a few times under partner's close scrutiny and criticisms.
6. Be seated and reach forward for imaginary objects. (If possible have equipment at hand so that demonstrations and practice can be realistic rather than imaginary.)
    a. An imaginary book.
    b. To work at an imaginary typewriter.
    c. To write at an imaginary desk.
    d. To hold the imaginary wheel of a car.
7. Remain seated and engage your partner in casual conversation, shifting your position to legs crossed (a) at ankles, (b) above knees and (c) at knees, noting which position or positions are pleasing and which unpleasing and which give the fewest tensions.

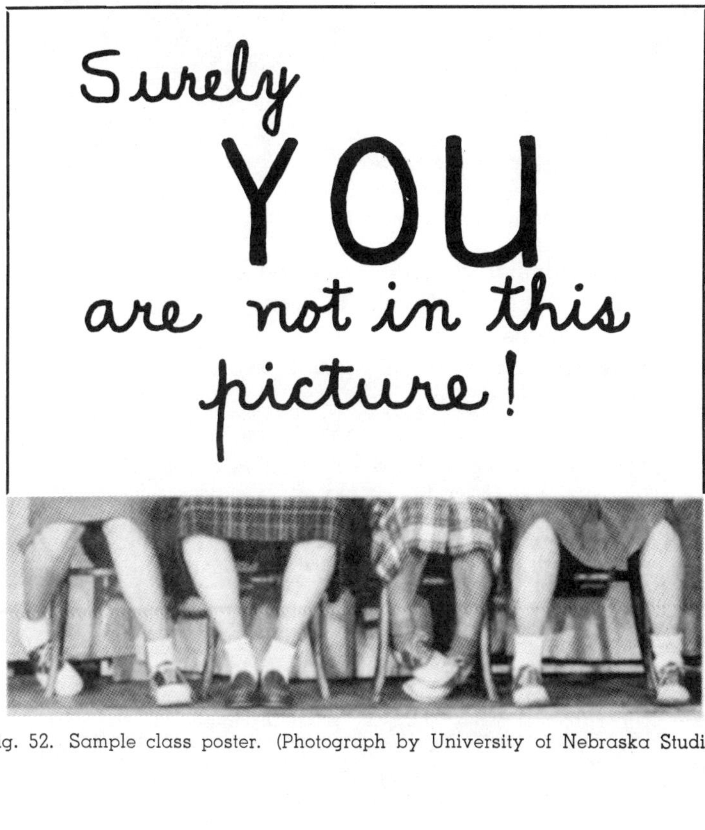

Fig. 52. Sample class poster. (Photograph by University of Nebraska Studio.)

Fig. 53. Sample class poster. (Photograph by University of Nebraska Studio.)

8. Repeat Number 7 placing the feet in various positions—front, right, and left—and make the same notations.

9. Sit with knees together (a) with feet also together and (b) with feet apart, noting which position is the better, both esthetically and mechanically.

10. Sit with knees spread apart (a) with feet together and (b) with feet apart, noting how unattractive are both positions.[27]

11. Practice sitting on the floor keeping the trunk as erect as possible. (Whenever the class is seated on the floor for a discussion period the teacher should insist on good sitting postures.)

---

[27]Exercise Numbers 7 to 10 are designated to give the feel and looks of the various right and wrong ways. These exercises will be more effective if mirrors are available. If they are not, urge the pupils to try these exercises before their mirrors at home, reviewing poses in Figures 50, 52, and 53.

a. With feet crossed in front and knees out.
b. With feet and knees together and swung to one side.
c. Study which is the better as far as correct body mechanics are concerned.[28]

Suggestions for Out-of-Class Practice

Students should be urged to observe and analyze the various sitting positions of people about them and practice those that are pleasing and effective. They should practice not only at lowering themselves onto and rising from the floor but also at lowering themselves into and rising from low seats of various types such as low, hard stools and low, soft cushioned davenports. Groups together should figure out what to do and what not to do and which ways are the most pleasing and at the same time most energy conserving. Then they should practice these agreed-upon, best ways.

Also it would be well for them to have group discussions and practice of that social situation which all too frequently is an awkward one for the inexperienced, namely, the act of being seated at a table while another slips the chair into place. The schoolgirl not only accepts this courtesy from her escort but in turn she performs this service for an older woman; therefore, both being seated and placing the chair for another should be practiced.

Similarly it will be helpful to many if time can be found for discussion and practice at correct mechanics of getting into and out of automobiles.

Proper Chairs and Worktables

A workchair used at a desk or table should have its seat adjusted to the height of the lower leg of the person who is to use the chair, the measurement taken from the center of the knee cap to the sole of the foot. The best type of workchair is narrow and has a wide opening at the back so that the hips can be thrust through the opening with the backrest actually

[28]Which position is the better at any given time will depend, of course, upon the clothing worn at the moment. A position that may be mechanically correct and at the same time pleasing when one is dressed in slacks or gymnasium costume may be unattractive when wearing a skirt, especially if the skirt is short and tight.

supporting the spine at the waist line. So called "orthopedic" chairs are of this type. The pressure on the spine is conducive to good posture, thus materially assisting the delay of fatigue.

The worktable that permits the best use of the body is of such a height that when a person is seated at it, the forearm can rest on it without having to lean down to it or without having to hunch up the shoulders to get up to it. The standard cardtable height is a good sitting-working height for the average woman.

STOOPING, LIFTING, AND CARRYING

Perhaps no principles of correct body mechanics are so universally violated as are those of stooping, lifting, and carrying. Yet there probably are no body movements which, if performed continuously incorrectly, will do so much damage to the body. For that reason we will go into the techniques of these movements in detail in the hopes that teachers will explain them carefully to their pupils and will drill them thoroughly in correct mechanics.

The Effective Use of the Body in Stooping

The vast majority of people perform most body movements "somewhat" correctly without special training (they do it "naturally," as we say) but it is indeed rare to find a person who performs stooping, in particular, correctly without special training and much practice. Out of a clear sky ask a group of people to stoop down to pick up some object and watch what happens. Practically everyone of them will flop over from the waist and reach down for the object with an extended arm reach, keeping the knees straight and hips pointing skyward. It will be a rare group that has even one person in it who will lower the body by bending at the hips and knees, thus lowering the hands down to the object and placing on the leg muscles, instead of the back, the task of returning the body to the upright position. If the stooping is to involve reaching for an object of noticeable weight to be lifted, it is all the more important that the position taken be one that will make the leg muscles save the back.

Mensendieck gives interesting comparative observations of Friesland and German peasants who do much field work requiring continual stooping. Friesland peasants stoop to their work by bending at the knee and hip joints and lowering their bodies to the task. As a group they have straight backs, strong abdominal walls and well-developed erect posture. On the other hand, German peasants habitually stoop to their work by keeping the legs straight and bending over from the waist and back, and reaching out from the shoulders. On the whole they have poor posture, broad hips, strained backs, and sagging abdomens.[29]

Pupils should have explained to them the general underlying principles of body mechanics which apply to stooping, lifting, and carrying. (See pages 146 and 148.)

Mechanics of Stooping

After discussing the basic principles the pupils should be ready for the following instructions:

1. Keep the body erect. Stand tall, pull in the abdomen, and tuck the hips under.
2. Stand close to the object to be contacted.
3. Have one foot in advance of the other with the weight between the feet. If stooping to lift a heavy weight which must be grasped with both hands, have the feet apart sideways with weight on both and squat down as close to the object as possible (Fig. 54).
4. Keeping the trunk erect, bend at the hips, knees, and ankles, lowering the body as far as is necessary to grasp the object. Do not reach down from the shoulders, thus bowing the spine, relaxing the shoulders and spreading the buttocks.
5. If the body proves to be so far away from the object that such a long reach is necessary as will disturb the body balance, shift the feet closer to the object.
6. Now the body is in a position both to lift the object and return the trunk to erect position by exertion of the leg muscles instead of the back muscles. In other words, the legs save the back. When the object to be lifted is not heavy or when there is no object to be lifted yet it is necessary to stoop down to the floor, a slight relaxation of the upper spine and shoulder girdle to permit a longer reach of the arms downward will save unnecessary expenditure of energy in lowering the body farther

[29]Mesendieck: op. cit., p. 97.

Fig. 54. Sample class poster. (Photograph by University of Nebraska Studio.)

than would otherwise be necessary. But there should be sufficient bending at the knees and hips to lower the trunk enough that this relaxation of the upper spine and shoulder girdle will be only slight.

If the knees creak, don't worry. They merely need lubricating, and exercising them thus will improve the circulation of blood and bring needed lubrication to them.

The Effective Use of the Body in Lifting

Much of work, be it in factory, home, or farm, requires heavy lifting. Even white-collar workers (teachers, office girls, and clerks) encounter many occasions when they must lift heavy objects or carry heavy suitcases and the like. Steinhaus warns that "because of structural weakness in the pelvic floor of women they should avoid jumping from heights, heavy lifting,

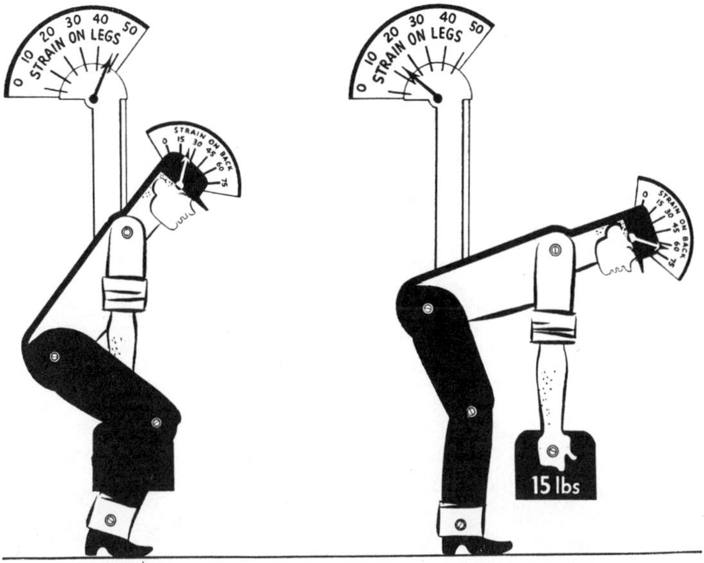

RIGHT WAY
Legs carry load; back strain minimized.

WRONG WAY
Legs doing little of work but back strain severe, as dial on head indicates.

Fig. 55. The mechanical man gives a lesson in weight lifting. (Courtesy of Bethlehem Steel Corporation.)

and other activities that greatly increase intra-abdominal pressure."[30] But heavy lifting cannot always be avoided, and therefore it is important that all girls and women learn proper techniques of lifting with the best mechanical use of their bodies in order to avoid back strains and sprains.

Back muscles were never intended to do the major work in heavy lifting, and they perform badly if driven to assume this task which the larger leg muscles should take over. Whether the back muscles or the leg muscles do the major part of the work depends entirely upon the position taken in the preliminary stoop which sets the mechanical pattern for the lift.

Bethlehem Steel safety engineers have developed a mechanical stooping-lifting man in order to demonstrate to their workers the difference in strains between the right and wrong way of lifting heavy objects. Figure 55 shows that when the

[30] A. H. Steinhaus: Fitness and How We May Attain It. J. Health & Phys. Educ. 14:456 (October) 1943.

stooping is done from the squat position the strain on the legs is 40 + pounds, and on the back is 20 + pounds, but when it is done from the bending forward position the strain on the legs is only 15 + pounds and on the back 60 pounds. As is so wisely said in safety campaigns "Use your head to save your back."

The teacher should explain to the class those basic principles of body mechanics that apply to lifting (see p. 148) and proceeding from there should guide them into an understanding of the rules that follow.

Mechanics of Lifting

1. Whenever possible to divide a heavy load into two or three lesser loads, do so and lift the parts one at a time.
2. If an unusually heavy load cannot be divided into two or more lesser loads, find some one to help you.
3. Stand close to the object to be lifted.
4. Regulate the depth of the preparatory stoop to the amount of weight to be lifted; the heavier the weight, the deeper the knee and hip bend to give the leg muscles added stretch for the stronger contraction that will be needed following the stoop to lift the heavier weight.
5. Grasp the object firmly and as near its center as possible, the type of grasp depending upon the size, weight, and shape of the object. Hold it as close to the body as possible so that its weight will be centered as much as possible over the base of support and near the body center of gravity.
6. Keep the weight evenly distributed between both feet.
7. Get set for the load. In lifting an object we unconsciously adjust our muscles to handle the load. The more skilled we become in judging the weight of a load, the better can we prepare the muscles to do the work that needs to be done and therefore the more skilled mechanically do we accomplish the work. Unexpected loads usually catch us off balance and readily "throw us" since we haven't "prepared" for them. Note the difference in catching a heavy object thrown at you expectedly and one thrown unexpectedly. In the first instance you brace yourself and muscles are contracted in preparation to receive and support the weight. If your advance estimate of the weight is good and you have sufficient strength to cope

with it, your efforts will be mechanically effective; but if you grossly misjudge its weight in either direction, you will be thrown off balance either because you have tensed too much or not enough.

8. Straighten the ankle, knee, and hip joints, thus pushing the body up into erect standing position.

9. The heavier the object the slower the movements. Use a slow, steady lift.

10. If the object is heavy enough to warrant it, maneuver it onto one knee so that the knee can help the arms support it.

11. If the heavy object is such that it can be hugged up to you, lift it onto one knee and then, as you straighten the legs, gradually move it up to a chest hold.

12. Set objects down by reversing the lifting process.

The Effective Use of the Body in Carrying

Whenever possible divide a load and thus make several trips of carrying it rather than carrying a too heavy load in one trip.[31] This technique is always employed by wilderness canoe parties who must portage frequently and, being far from civilization, cannot risk strains and sprains. It is, also, always employed by the wise worker at home. Strains and sprains acquired by unwise body mechanics are inexcusable under any circumstance. The rules for carrying vary with types of loads, but there are certain fundamental rules to be observed as given below.

Mechanics of Carrying

1. Keep as erect and tall as possible considering the load.

2. Keep the load as close to the body as possible and centered as much as possible over the pelvis.

3. When unable to keep the weight of the load directly over the body base keep it as close to the body as is possible and counterbalance the instability of the load by moving a portion of the body (usually an arm or the upper trunk) to the direction opposite the instability as much as is necessary to

---

[31]There are on the market these days various kinds of carts, wheelbarrows, and contrivances for women to use to make the carrying tasks easier in marketing, hanging out the wash, and gardening. Conservation of energy is always wise.

bring the weight (the body and the load as one weight unit) to center over the base of support.

4. Carry a load over head, up on one shoulder, or on one hip whenever possible, shifting from side to side as the muscles tire (Fig. 56). (Young mothers these days frequently carry their babies astride, first on one hip and then on the other. In this way the stronger muscles do the carrying work and the load is carried close to the body center of gravity making it seem less heavy; and, incidentally, the baby has an excellent saddle to ride on for its own comfort. This position is better for holding the baby while standing still than for walking since the load on the one hip interferes with natural gait.)

5. If possible do not carry a load directly in front of the body.

6. Do not carry a load with its weight hanging from the shoulder out away from the body.

7. When carrying a bucket or suitcase in one hand raise the opposite arm sidewards.

8. When carrying a load on the back keep its weight evenly distributed between the shoulders, keep the back flat, and bend forward a little from the hips.

9. When a load of noticeable weight is to be carried avoid carrying it in such a way that it raises the center of gravity of the body. In other words, don't let the load make you top heavy.

10. Regulate your rate of walk to the load. There is definite relation between the weight of a given load and the rate of walk in carrying it in reference to the most effective use of energy. Steindler of the University of Iowa says, "A man carrying a load of 45 to 55 pounds on the march conserves his energy best by walking at a velocity of 2.8 miles with a step length of 75 cm. and a cadence of 125 paces per minute." He goes on further to comment on the interesting fact "that the marching regulations of our Army, probably through a long period of trial and error, seem to have come to the same conclusions, and the prescribed step length, cadence, and marching velocity come surprisingly close to the figures of the theoretical kinesiologist.[32]

[32]A. Steindler: What has Biokinetics to Offer to the Physical Educator. J. Health & Phys. Educ. 13:509 (November) 1942.

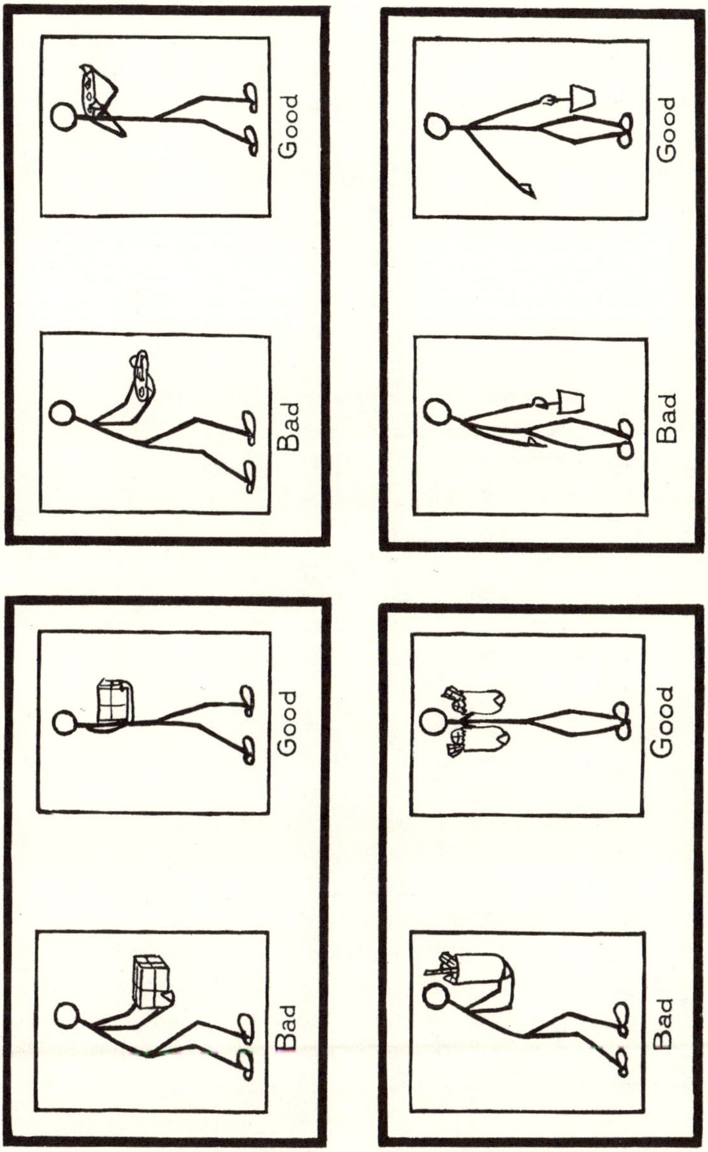

Fig. 56. Carrying. (Arranged from Purdue University, Department of Agriculture Extension: Keeping Fit for Farm Work.)

Practice Techniques

The teacher should, as in other phases of body mechanics, build up in her pupils an appreciation of correct stooping, lifting, and carrying through a discussion of basic principles. Then she should have demonstrations of the right and wrong ways of doing these things and should have pupils practice intensively at correct mechanics. Following are suggested class routines:

Practice Drills

1. Practice stooping without lifting.
2. Practice stooping and lifting an imaginary object.
3. Practice stooping, lifting and carrying
   a. Several books at a time.
   b. Suitcase loaded to usual travelling weight.
   c. Chair, stool, or bench.
   d. Rag doll the weight and size of a baby.
   e. Tray of dishes, both small and large.
   f. Bucket of water.
   g. Sandbags of weights varying from 20 to 50 pounds.

There will not be time for all to practice at each of the above but the class should be divided into groups with each group demonstrating some one of the above while the rest of the class criticizes their efforts.

4. Practice carrying some of these objects up a few steps.

Teaching Suggestions

For this work it is important to have actual equipment for the pupils to work with. They will never get the correct "muscle feel" in lifting and carrying imaginary objects. The only pieces of equipment suggested above that may be difficult to find are the rag doll and the sandbags. These the teacher can make at home and save them to use year after year. A little experimentation will reveal the patterns and sizes. The home economics teacher will probably be glad to assist with the rag doll pattern of actual body size.

For sandbags the following instructions will prove helpful:

1. Make of heavy denim stitched three times.
2. Make in varying sizes for the various weights desired.

3. Make one each to hold the following amounts of sand: 20, 30, 35, 40, 45, and 50 pounds.

4. Mark its weight on each bag.

5. Fill with sand that has been washed and thoroughly dried.

6. Have the bag large enough so that the sand does not have to be packed tightly.[33]

REACHING AND PUSHING

The basic principles of the correct mechanics of reaching and pushing may be stated as follows:

Fig. 57. Reaching.

1. The body segments should be maintained in correct alignment and the center of gravity of the body should be kept as much over the base of support as possible.

[33]Miss Ruby Holton has done interesting experimental work with sandbags at Drake University. This has been reported in her article in the Journal of Health and Physical Education, 13:473 (October) 1942.

2. The center of gravity of the body should be kept as close as possible to the object reached for.

3. Force should be applied through the center of the weight of the object and in the direction in which the object is to be moved.

4. The heavier or larger the object to be moved, the larger should be the base of support.

Mechanics of Reaching

After the teacher has discussed the basic principles above, she should instruct in special techniques of reaching, as follows:

1. In reaching up for an object keep the weight centered as much as possible under the object (Fig. 57).

2. Instead of reaching to the point of straining move closer to the object. If reaching far up use a stepladder or stool in order to move closer to it.

3. In using a paint brush, mop, or broom in an out-reaching movement, instead of reaching ever farther away as work progresses, move the body along, keeping close to the ever changing work, thus keeping the center of gravity close to the base of support.

Mechanics of Pushing

The following instructions should be given on the mechanics of pushing:

1. Push a heavy object rather than pull or carry it (Fig. 58).

2. Stand close to the object; the heavier, the closer.

3. Apply the force in the direction in which the object is to be moved.

4. Apply the force through the center of the weight of the object.

5. Have one foot forward with the weight on it ready for the free foot to advance beyond it as you start to walk forward propelling the object before you.

6. Lean a little forward to brace the hands against the object.

7. Keep the back straight and stand close.

Fig. 58. Pushing.

8. In pushing a very heavy object:
   a. Crouch down to it with one foot close to it and the other extended to the rear as a brace.
   b. Keep the hips down.
   c. Place the hands or shoulders on the object behind its center of weight.
   d. Keep the back straight.
   e. Apply force and walk forward with a long, crouching stride, pushing the object ahead of you with the legs doing most of the work.

Teaching Suggestions

It is doubtful if there is need to drill pupils in the correct mechanics of reaching but since many women have at times such an inordinate urge for moving furniture about, it is well to teach girls the proper mechanics of pushing heavy objects. After the fundamentals have been explained it would be well to demonstrate, then have groups take turns at moving some heavy object about the gymnasium room. A piano is usually at hand and it can be worked at by a fairly large group at a time. It should not be shoved around carelessly. No damage need come to it if used only a few times a year for such class practice.

RELAXATION

If we could learn "how to balance rest against effort, calmness against strain, quiet against turmoil" we would thereby be able to "achieve some degree of peace" in this life, to "offset hypertension in personal relationships," and to "untangle some of the taut threads in life."[34] In other words, we need to relax, but since most of us do not relax readily and naturally, we need to learn how to relax.

The word "relaxation" is frequently used to express three different ideas as follows:

1. As most people use the word they mean merely diversion such as change from work to play or change from one type of work to another type even though the thing one changes to may not of itself be relaxing. As they use the term, it refers to rest from certain mental and physical processes while engaging in other activities. The tensions or fatigue which this sort of relaxation releases is a false fatigue—it is merely boredom, monotony, or sheer need of exercise.

2. As most physical educators and sports trainers use the term they mean release from muscular tensions even though the means of procuring that release are active exercises calling forth action of a different set of muscles—yet nevertheless action.

3. As the physiologist uses the term he means complete inaction. We will in this discussion take up the last two meanings and for the sake of clear distinction between the two we will speak of the first as "active" relaxation and of the other as "passive" relaxation.

Values of Relaxation

The heart relaxes automatically between beats and thus restores itself without conscious effort. The person who can completely relax any muscle group whenever it is not actively needed for some body activity is the person who keeps in excellent mettle as he goes along. He is the person who can stay on the job and get the world's work done for he isn't wearing out needlessly. The body must have opportunity to re-

---

[34] Josephine Rathbone: Relaxation. New York, Teachers College, Columbia University, 1943, p. 90.

cuperate and retrieve its energy loss, and relaxation is the means of doing this. It may take three forms (1) sleep at night (which is not always relaxing and hence not always recuperative), (2) brief rest periods during the day and (3) constant relaxing of unneeded parts throughout the day. The more a person can relax unneeded parts as he works and plays, the longer he can keep at a task, the longer he can delay fatigue, and the less does he need special rest periods to recover. Also the more he can completely "let go" when he rests and sleeps, the more refreshed is he after the rest and sleep, and the less time does he need to give to it. If relaxation is total, the body is completely inactive. The person who "can't let go" uses up energy on muscles contracted to no purpose. Overtiredness keeps muscles in a state of contraction and the unhappy victim cannot rest well.

The ability to relax increases efficiency in both work and play. In many emergencies a person must work long hours under strain; he who can relax holds out the better. To relax hourly, daily, weekly on the job and keep going effectively is better than to push at tension for a long time and then stop completely for a while to recover. The skill to relax at will is one of the important techniques in the proper use of the body for effective living.

Physical Tensions

According to Rathbone pressures to get things done more quickly, or in larger doses than usual, or to excel or outdo others most probably cause most of our physical tensions. The tense person is cramped and restricted in his movements and is ungraceful. He is apt to be irritable, distraught, and touchy. He fidgets, taps his fingers, holds parts of his body cramped or holds parts in queer positions. He hunches his shoulders. He has restricted breathing from loss of flexibility of the chest and he probably suffers from some maladjustment of the breathing mechanism. Most of all, his tenseness shows up in the spine, causing discomfort in the internal organs.[35] Also his circulatory system has unnecessary demands made

[35]Rathbone: op. cit., pp. 13-20.

upon it to supply oxygen to tensed parts that should be relaxed.[36]

The Release of Tensions

To release tensions in the manner indicated by most people in the use of the term "relaxation" means to do something active. This is quite different from the interpretation Jacobson[37] places on the term "passive relaxation" when he means doing nothing, absolutely nothing. For the sake of pupil training in the art of release of tensions, exercises suggested both by the "active" relaxation exponent[38] and techniques of "passive" relaxation will be presented in the material that follows.

Exercises for Release of Tensions

Rathbone maintains that tensions can be reduced by (1) rhythmic exercises (2) flexibility exercises, particularly for the spine, chest, and diaphragm and (3) concentrating on the reduction of tonic contractions in localized muscles.[39] The latter method is the passive method to be discussed later. To procure substantial relief from tensions, particularly in the spine, takes hours of active practice at relaxation techniques. Relief and re-education do not come merely by giving up some of the causes of tension.

Much re-education of the tense person is necessary to avoid tensions. He probably has acquired, for example, the bad habit of holding his breath whenever he works intently. This tenses the diaphragm and if it is frequently and continuously repeated the diaphragm tends to lose its flexibility.

Exercise Routines

Flowing, swinging movements and stretching movements make up the best types of exercises for the release of tensions.

---

[36]Josephine Rathbone: Relaxation in the Emergency. J. Health & Phys. Educ. 14:7 (January) 1943.

[37]Edmund Jacobson, M.D., noted physiologist and authority on the subject of relaxation, has devoted many years to intensive research on this subject.

[38]The two groups are not in disagreement as they refer to two entirely different activities each with its own merits for its own purposes; they are merely unfortunate in their use of the same term, "relaxation," for their two different types of activities.

[39]Rathbone: Relaxation. Op. cit., p. 13.

A suggested list for the teacher to give her class is as follows:

1. "Elephant Swing." Stand with feet apart, let the head, arms, and trunk hang heavily downward and forward. Swing the trunk from side to side letting the arms "flop along" as if they were an elephant's trunk.

2. Leg Swinging. Hold the back of a chair or brace against a wall, swing the outer leg forward and backward, letting the leg be heavy and keeping the hip joint loose. Change to the other side. Let the leg swing loosely "any old way."

3. "Lazy Rolling." Lying on the back with arms stretched far overhead, raise one hip and try to swing it across the body letting the rest of the body follow heavily until you are turned over on your face.

4. Head Circling. Bend the head forward and swing the chin towards the right shoulder; slowly raise the head upward and, keeping the chin pointed toward the right shoulder, let the head fall back as far as possible; swing the chin across to point behind the left shoulder; start lowering the head to let the chin point in front of the left shoulder; keeping the head down, swing the chin back to center. Keep the movement slow, smooth, and flowing, describing a circle with the head. Make circles both to the left and to the right.

5. Release of Tension in Chest. Lie on the back with the knees bent and feet drawn up on the floor near the hips. Breathe deeply and slowly clear down into the abdominal cavity. Open the mouth and let the air out.

6. Release of Tension in Diaphragm. Sit erect either on a chair or on the floor. Spread the hands out over the ribs at the sides. Breathe deeply and slowly, gradually increasing the outward swing of the ribs. Each time as you reach the greatest swell of the diaphragm hold the breath momentarily and then quickly force all the air out, letting the escaping air make a great noise.

7. Stretching. Lying on the floor on the back, bend the knees and bring the feet up near the hips to relieve the pull on the lower spine. Raise arms over head. Stretch and twist in all directions. This should be done for a few moments every day.

8. Brief rest moments, lying on the back on the hard floor. Rathbone maintains that the best position for resting is on the

back with the legs turned outward, the knees bent slightly and the arms down at the sides of the body but not touching it and with elbows partially bent and turned out. This position relaxes all "ligamentous strain."[40]

Techniques of Passive Relaxation

According to Jacobson we must learn to relax the muscles of the eyes, throat, and face as much as those of the big-muscle groups. He speaks of the eye and speech muscles as the sense key muscles and calls attention to the fact that when we have imaginary conversations with ourselves the throat muscles tense and go through certain motions of speech, and when we imagine objects the eyes are tensed to see them inwardly. For the mind to be as blank as possible, it is necessary therefore to relax completely. To make the mind blank we must "turn off" these mental activity muscles. If we can keep the eye and speech muscles relaxed for as much as 30 seconds, we will be pretty sure to fall asleep by then.[41]

Practice Drills

Jacobson gives six steps in the fundamental techniques of acquiring the ability to relax.[42] Four of these are outlined for group work as follows:

---

[40]Rathbone: Relaxation. Op. cit., p. 210.

[41]Jacobson: You Can Sleep Well. New York, McGraw-Hill Book Co., 1938, pp. 230-38.

[42]This material is the authors' own adaptation into class drill form of the material offered in "You Can Sleep Well," previously mentioned. The material in brackets is the authors' suggestion for class adaptation to homework. This arrangement into outline form for class group and home instruction has received the approval of Doctor Jacobson. However, no teacher should undertake this classwork without first studying the instructions in the original book. The material presented here is merely suggested for class outline work. During the war Jacobson's techniques were adapted for instructional use by the Naval Pre-Flight program. The trainees were supposed to work at these techniques for an hour a day over a given period of time. A report from Dr. C. H. McCloy of the University of Iowa who, as Civilian Consultant to the Navy, prepared the course of instruction under Dr. Jacobson's supervision, informs that these men who took the course broke all records for studies and well-being. For those students who wish to practice relaxation techniques seriously, the practice suggestions that follow should be posted for easy reference.

A. Arm Relaxation

1. Lie on the back, close the eyes. Do not cross the legs.
   a. Raise the right arm vertically with the fist clenched. Note the muscle feeling in the whole arm.
   b. Let the arm fall of its own dead weight and let the fingers uncurl. (Watch out for wrist watches.) Let your arms fall like the arms of a rag doll no longer being held up by its mistress.
   c. Let the arm rest as it fell, do not move it, but do not forcibly keep it still. Just let it go. "Relaxing is **not doing**."
   d. Repeat a to c three or four times.
2. Same, using the left arm.

3. Practice getting tensions in the arms without raising them or clenching the fists. Then "let go" gradually without moving them, but be careful not to hold them still.

4. Practice this tensing and "letting go" several times as you lie quietly with arms at side.

5. Sit up a few moments and then again lie on the back without the legs crossed and focus your attention on the tensions in the arms that exist without consciously tensing, and practice "letting go" from there.

[Explain to the class that what they do in class is not really practicing to go to sleep because we can't devote the time to let them actually do it. But we are showing them how to practice when they do go to bed. In that situation they are to do it the first night a few times with right arm only, the second night a few times with left arm only, the third night a few times with both. Each time they are to lie as quietly as possible after the arm falls to the side, letting go and refusing to permit themselves to shift about. Then the fourth night, and on indefinitely, they are not to tense the arms to start but start relaxing from whatever tension exists in the arms as they lie at their sides. They should do this each night for a long time until they can quickly relax the arms at will. They should not force any part to be still because forcing is not relaxing. "Relaxing is **not doing**." When they catch themselves doing this, they should tense this part consciously and then "let go."]

B. Leg Relaxation

1. Remove the shoes and lie on the back with the legs side by side and the eyes closed.

   a. Keeping the leg straight, stretch the toes as far away from the rest of the body as possible and hold this position for a few moments. If you are subject to cramping in either foot let go at once and instead of stretching the toes away from the body, bend at the ankles and stretch the toes towards the knee, pushing the heels away from the rest of the body. Use this method of tensing instead of the other. If you did the other and got a cramp, rub the painful part for a while and then get up and rise on your toes and sink down on the heels a few times.
   b. "Let go" gradually through the whole length of the leg down to the toes, taking one or two minutes to it.
   c. Remain relaxed a moment.
   d. Repeat a few times (1) stretching (or tensing), (2) letting go gradually and (3) remaining relaxed.
   e. Repeat d except let go suddenly throughout the entire leg and foot.

2. Combine relaxation of both the arms and the legs at same time, but do not tense to start, just begin with whatever tensions exist. Try a few times, observing parts 1, 2, and 3 carefully, tensing, letting go, and remaining quiet.

3. Work in couples. The first girl assumes the lying position and the second girl works with her. The first girl is told to relax.

   a. The second girl puts a hand under the first girl's heels and lifts her leg and releases it. If the first girl resists she is not relaxed; she is not relaxed also if her leg does not fall as a dead weight. Try again and again until the second girl feels that the leg of the first girl is relaxed.
   b. Exchange places and the first girl works on the second girl.

[If any girl has difficulty in relaxing as shown by this practice, she should practice it on her bed at night with someone to help her, and should keep this up as many nights as necessary until her assistant reports success. Tell the class to practice this for a few minutes each night when they first lie down to sleep and to keep it up for many nights.]

C. Relaxing the Chest

1. Lie on the back.

2. Breathe in a little more air than you usually do, but not a deep breath, and do not force it out, just let it come out. Do this only two or three times, noting the different muscular sensations in your chest as you breathe in and out.

3. Let the chest go, as you let your arms and legs go in previous practice. Breathing will be shallow when relaxed. Work at getting this feeling of relaxation, but do not do anything active—just let go and **not do.**

Tell the class that if they do not fall asleep quickly after going to bed to practice this for a few minutes and to combine it with relaxation of the arms and legs.

D. Relaxing the Speech Muscles

1. Lie in any position you like but with the eyes closed.
   a. Count aloud to 10 (1) noting the feeling in all parts of the speech apparatus and then (2) relax these parts for three or four minutes.
   b. Count again but this time to yourself and in such a way that anyone watching you would not know that you are counting. (1) Note the feeling in the tongue, lip, jaw, throat, and chest and (2) relax, being careful not to talk to yourself.

2. Repeat all of Number 1 a few times.

3. Repeat Number 1, combining relaxation of the entire body. Practice a few times.

4. Practice relaxation of the speech muscles without counting first—just relax from whatever tensions exist as you lie down and are not talking aloud.

5. Repeat Number 4, relaxing the entire body at the same time.

[The class should practice this briefly each night after going to bed and should try to acquire the habit of not talking to themselves when it is time to go to sleep.]

SPECIAL HINTS FOR SPECIAL JOBS

In Homemaking

House and garden work furnish lots of exercise but it depends upon how they are done whether they add to physical fitness or give strains and nerve drains. If the homemaker does the work in the right way and dresses appropriately so that the muscles can have free play, she will not have to "join a gym class" to keep fit: She already belongs to a wonderful "keep fit" class which meets daily, rain or shine, year in, year out. She is the star member of the class and can set the pace.

The secret of doing house and garden work correctly from a body mechanics point of view is observing the proper mechanics of standing, walking, stooping, climbing, lifting, pulling, pushing, and reaching (Fig. 59). These activities just about make up the sum total of the whole job. The teacher should lead her pupils in a discussion of the mechanical principles relating to the usual run of homemaking activities and then should take up with them the following specific instructions, having them interpret each in terms of fundamental principles:

1. When bending over an ironing board, stove, or sewing table keep the back straight, one foot forward, bend only from the hips: Don't curl forward through the spine.

2. Whenever possible use long-handled utensils to prevent the need of bending over at your work or else very short-handled ones and stoop to a one-knee kneel position.

3. When hanging out clothes either kneel down to the basket whenever you pick up more clothes or place it on a high rest so you will not have to stoop down to it: Don't bend over to it from the waist and spine with arms and shoulders reaching down to it.

4. In sweeping, swing the broom with hip action. Use the shoulders to guide it, not to push it.

5. In reaching up, either to hold some object up high or to grasp it to pull it down, have one foot well in advance for a broad base and move up to the object as closely as possible without forcing the head to be tipped back uncomfortably in order to look up at the object. In reaching, the body should follow a straight diagonal line from the rear foot to the outstretched fingers, with the head thrown back but little out of

Fig. 59. Effective use of the body in homemaking tasks. (Purdue University, Department of Agriculture Extension: Keeping Fit for Farm Work.)

this line in order to see the object above. This same position is the correct one when reaching merely to dust or to wash a high spot on the wall. (Review Fig. 57.)

6. When working on hands and knees the best mechanical position is with the arms and legs at right angles to the floor and the back parallel to the floor. Keep the body weight back

on the hips, not forward on the shoulder girdle. The working movement should come from the hip joint except that which is necessary for the arms as they swing from the shoulder to permit the hands to perform their work.

7. Have sinks and worktables high enough that you need to bend over them but very little. Don't let a man who doesn't do much work at a sink tell you what height it should be. Worktable heights can be remedied easily by adding lifts to the legs. Practically no kitchen worktables on the market have sufficiently long legs. The added lift saves much back strain.[43]

8. The ironing board should not be so low that you have to bend down over it but, on the other hand, it should not be so high that it causes a hunching up of the shoulders in order to put pressure on the iron.

9. When you need to stoop over to put things in the low oven, pick things up off the floor, use a low-handled dust pan, or help little children, you should go down on one knee to the task, keeping your trunk erect. If you bend from the hips and spine keeping the knees straight you will make drudgery of your work; you will be apt to have frequent backaches, and you will acquire wide, spreading hips. Make the powerful leg muscles do the hard work instead of the back muscles.

10. For good health, good looks, and the joy of living, always stand and sit tall, pull the abdomen in, and keep the hips tucked under.

In Industry

During the past war it was the privilege of one of the authors to serve as a member of the Chief of Staff's National Civilian Advisory Committee for the Women's Army Corps and in that capacity she had several opportunities to observe, first hand, women on the job in Army camps, in hospitals, and in industry, and to talk with their personnel advisers. The problem of woman's fitness for her job and of her ability and capacity to do it correctly was a serious problem during the war.[44] Al-

---

[43] When planning the places about the house and garage to keep work tools remember that it requires less energy to reach for them than to stoop for them.

[44] Augusta Clawson: Shipyard Diary of a Woman Welder, New York, Penguin Books, Inc., 1944, gives a splendid account of how women fitted themselves into the ship-building program.

though the American women served loyally and surprisingly effectively in industry and in the armed services, the war effort would have been even more successful if there had not been the general lack of fundamental training in correct body mechanics. There were many important body-mechanics problems yet unsolved when the war came to an end.

Women as a rule are not familiar with the type of work met with in industry and therefore need careful instruction and drill in its techniques. They are a greater risk than men because (1) they have the bad habit of wearing high-heeled shoes which give them poor balance, (2) they do not have the proportionate strength of men,[45] (3) they have a high percentage of menstrual period handicaps,[46] (4) their work habits are formed around work of quite different techniques than those of industrial jobs and (5) they do not tend as much as do men to study out for themselves the best ways of doing things to conserve strength and energy.

In view of this lack of knowledge about the effective use of the body in activities of industry and in view of the fact that so large a number of girls are graduated from our schools into industrial jobs, the physical education teacher should see that all girls leave school with bodies fit and well skilled in the fundamentals of body mechanics. In addition, they should be given a thorough understanding of the basic principles of body mechanics so that they can solve their own body-mechanics problems as they arise later in life.

The following few fundamental techniques of handling tools should be taught girls and women, at the same time leading them to an interpretation of each technique in terms of basic principles:

1. Grasp the tool firmly but keep relaxed.

2. Put the body weight back of the tool whenever the work permits it.

3. Keep the shoulders down and relaxed as you handle your tools: Don't get hunched up.

4. When necessary to apply pressure on a job place the

[45]Women have only 60 to 70 per cent as much muscle for body mass as do men.

[46]How industry met this difficulty is discussed in Chapter V under a discussion of the Billig exercises.

Fig. 60. Sample class poster. (Photograph by University of Nebraska Studio.)

Fig. 61. Sample class poster. (Photograph by University of Nebraska Studio.)

body in the line along which the pressure is to be applied and, if possible, exert the pressure at waist height. Stand a little away from the object with feet apart and one advanced (the left if you are right-handed) and leaning from the rear foot, slant the body toward the object. Keep the back straight.

For Charm, Poise, and Efficiency

There is much the teacher can do to help girls attain poise and charm as well as efficiency in handling the body. Important suggestions to pupils are as follows:

1. Movements that are performed efficiently are not only conserving of energy but also graceful (Figs. 60 and 61).

2. For grace of movement move slowly and smoothly. Haste tends to make jerky movements. If you must hurry, try to do it without seeming to. Practice moving to waltz rhythm and being deliberate in all movements. The studied and unhurried movements of the leading actresses are interesting to observe in these respects.

3. Perhaps no pose betrays your worst, off-guard self so much as your sitting posture. Nothing so readily marks the "country bumpkin," the "awkward Jane" as the knees straddled apart; nothing more plainly symbolizes the abandonment of youth than the habit of sprawling over a chair; and nothing more loudly proclaims the too intense individual than the habit of sitting clutching one's self with arms tensely folded on the chest or wrapped around each other. Keep your hands off yourself; clutching yourself makes you look years older. Let the hands rest relaxed in your lap and sit tall with feet together.

4. The correct posture for social dancing is the same as for standing and walking except that for the woman the arms are raised to permit her partner to hold her in dance position. This arm raising can readily result in hunched-up shoulders if it is not guarded against. The weight should be carried squarely over the pelvis and all features of correct body alignment should be maintained if one wishes to appear at her best on the dance floor. It is the ungraceful dancer who assumes a forward or backward leaning posture or who permits herself to fall into positions in which she is off balance.

MEMORY JOGGERS FOR BEGINNERS

STANDING

Do

Keep relaxed.
Stand tall and light as a feather.
Keep chin in and up.
Let the shoulders hang wide.
Keep buttocks tucked under.
Keep the abdomen pulled in.
Keep the knees loose.
Keep the sternum up.
For short periods, keep feet together, weight on both.
For long periods, keep one foot forward and carrying the weight.
Keep toes pressed to floor with thigh rotated outward.

Don't

Have arms akimbo.
Have feet apart sideways.
Have one hand on hip.
Be tense.
Be a "droop."

WALKING

Do

Stand erect.
Keep knees close together.
Move slowly and smoothly.
Toe forward.
Keep knees relaxed.
Seem to glide.

Don't

Move fast.
Pound the floor.
Rear back from hips.
Lean forward.
Mince along.
Be a strider.
Swing arms far.

CLIMBING STAIRS

Do

Stand erect.
Use whole foot.

Don't

Lean forward.

RUNNING

Do

Keep trunk erect.
Swing leg from hip joint.
Keep knees, ankles "soft."
Lean a little forward from hips.

Don't

Swing shoulders.

SITTING

Do

Sit erect.
Sit "firm astern."
Have hips clear back on seat.
Keep hands relaxed in lap.
Have seat adjusted to your leg length.

Don't

Straddle the knees apart.
Cross the legs at the knees.
Be a "tail sitter."
Clutch yourself with your hands.

STOOPING

Do	Don't
Keep erect.	Bend over from waist.
Bend at hip and knee joints.	Keep knees straight.
Keep hips in under.	Thrust hips up behind.

LIFTING AND CARRYING

Do	Don't
Keep erect.	Make the back do the work.
Let the legs do the work.	Carry from the shoulders and arms.
Keep the object close to you.	
Use the leg and back muscles.	
Keep the weight of object centered over the pelvis.	

REACHING

Do	Don't
Reach from hips.	Reach from shoulders.
Move close to the object.	

PUSHING

Do	Don't
Keep the back straight.	Push with arms and shoulders.
Apply force through the center of the weight of the object.	
Stand close to the object.	
Have one foot forward.	
Lean forward a little.	
If heavy, crouch to it.	
Push with back and leg muscles.	

RELAXING

Do	Don't
Let go.	Keep muscles contracted.
Turn off the mind.	Fidget.
Do nothing.	Talk to yourself.
	Hold the body in queer positions.
	Hunch the shoulders.
	Hold the breath.

PASSPORT TO EFFECTUAL LIVING

Charm, poise and efficiency! Passports to effectual living. In "What Every Woman Knows," Barrie's Maggie speaks of charm as "a sort of bloom on a woman." "If you have it," she says, "you don't need to have anything else; and if you don't have it, it doesn't much matter what else you have." Today's Maggie surely recognizes poise and efficiency as important accompaniments of charm. Poise (that inner calmness), efficiency (that adroitness and expertness of performance), and charm (that outward manifestation of a noble spirit)—all three are of great importance! There is much one can consciously do to acquire or intensify these attributes and this much each teacher should help her pupils achieve towards richer, fuller living.

CHAPTER VIII

HYGIENE TOPICS SPECIFICALLY RELATED TO BODY MECHANICS AND CONDITIONING

Many aspects of hygienic living are associated with the hygiene of exercise. Pupils should be instructed in the proper observance of those hygiene precepts which concern the effective use of the body. The major topics which should be discussed are taken up briefly in the material that follows. This information should be woven into the course in the fundamentals of body mechanics and conditioning.

BACKACHE

The best insurance against backache is the proper use of the body supplemented by back muscles that are sufficiently strong to do their work easily, and the best way to get strong back muscles is to exercise or work them into strength and endurance. The next best insurance against backache is to avoid activities that put too great a strain on the back, such as lifting objects that are too heavy or lifting heavy objects without throwing the major responsibility of the work on the powerful leg muscles.

Back Strength for Girls

Girls do not develop strong back muscles by being mollycoddled in their physical activities. Most of our school physical education programs for girls are too mild to give real physical fitness. As far as back strength is concerned, even the most vigorous sports for girls do not usually develop the back muscles sufficiently, and special exercises are necessary, particularly for the lower back. Hip Raising is suggested in Chapter IV for the lower back strength builder of the conditioning series. Many people will object to its inclusion in a series of exercises for girls. These are the people who, deploring the weak lower back muscles of most girls, are afraid to let girls do any exercises to build strength for those muscles. There are very few exercises that really engage the lower back muscles enough for conditioning. Hip Raising as used in the suggested conditioning series is such an exercise and a good one. Care, however, should be taken that the abdominal muscles are being strengthened equally, otherwise the stronger pull from the back will produce a hollow back.

Sacro-iliac Strains

Sacro-iliac strain (a dull aching due to strain of the ligaments holding the sacrum and the ilia together) is the most common form of lower backache in girls and women. Much of it comes, as previously stated, from weak musculature, but ligaments that do not have enough give to them are also to blame. Exercises for the stretching of ligaments that are too tight have been discussed in Chapter V.

During World War II many women workers in shipyards suddenly found themselves minus their old sacro-iliac pains and aching muscles thanks to the great amount of climbing up and down steep ladders which they had to do. This plus the general tough physical exercise, hours and hours of it each day, wrought miracles for them in terms of physical fitness. One woman who worked in the Portland Shipyards puts it this way:

> All that climbing does give one a good work-out. Wasn't I complaining about an aching sacro-iliac a few days back? Well, the aches seem to be gone forever. My muscles have tightened up so tremendously from all this exercise that the silly old sacro hasn't a chance of slipping anymore.[1]

Causes of Sacro-Iliac Pain

Much of the sacro-iliac pain which women complain about occurs only during the menstrual period, but the physical education teachers need be concerned chiefly for the sacro-iliac pains which occur without reference to the periodic functioning.

[1]Augusta Clawson: Shipyard Diary of a Woman Welder. New York, Penguin Books, Inc., 1944, p. 147. Miss Clawson was sent by the U. S. Office of Education to work in the shipyards incognito in order to study first hand the training needs of women shipyard workers.

If the pain is persistent, the pupil should be advised to see her physician. When the pain is severe it can seriously incapacitate the victim. In such cases complete rest and hot, wet applications and perhaps even taping of the joints are called for. Exercise must be avoided during this time, particularly trunk bendings and lifting of even moderately heavy objects. There are many causes of sacro-iliac strain such as (1) incorrect landing when jumping from great heights or even from moderate heights, (2) falls involving heavy sitting on the end of the spine, (3) slipping with a sudden throwing of one leg upward while the other frantically hugs the floor or icy pavement in an effort to keep from falling and (4) lifting objects that are too heavy or lifting heavy objects too frequently. All such emergencies can so readily occur in physical education classes if the teacher, as well as the pupils, is not on guard.

BATHING

One of the most valuable parts of an exercise period is the shower at the end of the "work-out." It gives the skin its conditioning exercise, toning it up, besides removing perspiration (which we hope the classwork was strenuous enough to produce) and toning the body in general. If the school does not have showers the teacher should exert all possible influence in trying to procure them since they are important class laboratory equipment. If the school does have them each pupil should for the final part of each class period take a warm shower followed by a cool or a cold one.[2]

Over a generation ago President Thomas of Bryn Mawr College warned her pupils of the social gulf that is bound to exist between people who take a daily bath and those who do not. For the sake of their social as well as physical training, the teacher should motivate her pupils to finish each physical education class hour with a shower.

FATIGUE

The problem of fatigue in relation to exercise has been discussed in Chapter II. The teacher should be familiar with signs

---

[2]For a discussion of showers, their construction, equipment, and rules for their use see Mabel Lee: The Conduct of Physical Education. New York, A. S. Barnes & Co., 1937, Chapters IV-VII and XI-XIV.

of fatigue in her pupils and should herself be convinced of the need of building strength and endurance as a preventive against fatigue. Beyond this she should refer to the physician those children who tire too readily; it is unwise to push children too hard against present capacity.

CONSTIPATION

Constipation is frequently spoken of as the great American habit since the people of no other nation spend such fabulous annual sums on cathartics. In view of this seeming widespread lack of knowledge concerning the hygiene of the digestive tract the schools should instruct all pupils in the causes, prevention, and cure of constipation. The teacher of physical education has a fitting opportunity to give this instruction when she is teaching specific exercises for strengthening the abdominal muscles. These exercises should be given a place in every physical education program. Such exercises and their relation to the relief of constipation were discussed in Chapter V.

A brief explanation of the meaning of the term "constipation" should be given to all pupils and be followed by a brief but carefully planned discussion of the causes of constipation, its prevention and cure,[3] with careful explanation of the part exercise plays in its prevention and cure. The special exercises should be carefully demonstrated and explained and the pupils should learn to do them correctly and then should be advised to do at least one of them daily whenever there is a tendency towards constipation.

MENSTRUAL FUNCTIONING

The Establishment of Correct Attitudes

Some time during the course in the fundamentals of body mechanics and conditioning the physical education teacher should instruct the girls briefly in regard to menstruation, its purpose, normal functioning, and hygiene, leading to an understanding of its relation to exercise.[4] This instruction should center around the establishment of correct attitudes towards

---

[3]Any good hygiene text will supply this information.

[4]Any good textbook on hygiene will give the necessary material for this talk. Safeguarding the pupil in her exercise program during the menstrual period is discussed in Chapter IX.

the menstrual functioning for, as Dr. Emil Novak of the Department of Gynecology of Johns Hopkins Medical School, has said:

> The role of the psyche in the production of primary dysmenorrhea is unquestionably a most important one, . . . The best prophylactic against this type of dysmenorrhea is proper training of the girl, preferably by her mother, as to what menstruation signifies and what it does not signify and, above all, that the function is not characteristically associated with pain and that it should interfere but little with the life activities of the girl. Unfortunately this ideal training is not available to most young girls in the puberal and adolescent years. Many mothers are loath to discuss these problems with their daughters and only too often are themselves ignorant of simple facts about the menstrual function. If any advice is given, it is far more apt to consist of admonitions tending to make the girl coddle herself during menstruation. Thus the impressionable youngster is apt to look on menstruation as a time when she should actually be unwell, and a step is taken toward the development of primary dysmenorrhea.[5]

The Woman Physical Education Teacher the Logical Instructor

Since mothers in such large numbers fail in this educational responsibility toward their daughters, it behooves the schools to assume this duty; and since the classwork of the woman physical education teacher is affected by the menstrual functioning of her pupils, as is the classwork of no other teacher, it is she who is usually considered the logical person to give this instruction.

If a life-sized manikin is available she should make use of it, permitting the pupils themselves to take out and replace the pelvic organs in their proper positions. This never fails to hold the deep attention of any group of girls. Questions should be solicited and if the teacher is not sure of the answers she should acknowledge this and ask the class for time to procure the correct answers for a later class report. Any question on this topic which a girl is interested enough to ask is worth the teacher's attention. There probably is no subject within the realm of anatomy, physiology, and hygiene about which more misinformation is passed along from child to child than the subject of the menstrual functioning and the reproductive system. If the teacher can substitute scientific facts for misinformaion and healthy attitudes for fears and taboos, her time will have been well spent. She should, however, avoid the

[5]E. Novak: Gynecologic Problems of Adolescence. J. A. M. A., 117:1950 (Dec. 6) 1941.

use of too scientific terminology, yet at the same time she should teach the use of correct terminology. To save much class time it is well to give the pupils in advance a list of carefully chosen readings on this topic. This is sure to be one topic all will follow through article by article.[6]

MUSCLE SORENESS

There is nothing to be gained by developing extreme muscle soreness although some muscular stiffness is unavoidable when muscles unaccustomed to being used are put into action. For information and advice to give pupils on this topic see Chapter II.

SHOES

There is no article of today's wearing apparel that so materially affects posture and one's capacity to use the feet correctly as does the shoe. If the shoe is too narrow or too short the transverse arch is crowded and cannot function: It becomes weak from disuse or incorrect use and in extreme cases it even breaks down. A weak or flattened transverse arch is shown by a callus that forms on the ball of the foot which may become hard and painful.

If the shoe is too narrow and the inner border swings too much away from a straight line it pushes the great toe and its tendons out of place, resulting in the development of a prominence called a bunion, which may become painful and even crippling. Also the throwing of the great toe out of place throws out of line the attachments of the ligaments supporting the longitudinal arch and that arch becomes strained and weakened.

If the heel is too high it causes an increase in the curve of the lower spine and tilts the pelvis downward. In many people this causes strain and backache. The higher the heel the more the knees must bend or the more the curve of the lower spine

[6]A few such references are:

1. Banning, Margaret Culkin: The Case for Chastity. Reader's Digest, 31:1 (August) 1937.

2. Diehl, Harold: Normal Sex Life. In Textbook of Healthful Living. 3rd ed. New York, McGraw-Hill Book Co., Inc., 1945, Chap. XVIII.

3. Meredith, F. L.: Reproduction and Sex. In Twelve Hours of Hygiene. Philadelphia, P. Blakiston's Son and Co., Inc., 1935, Chapter XII.

must increase its curve in order to throw the upper part of the body backward sufficiently to compensate for the sharp pitching forward of the body from the ankles. This leaning backwards from the hips (which Dr. Louis Collin called "hanging on the Y ligament") puts severe strain on the tissues over the front of the hips and markedly stretches the abdominal muscles until, if this stretching is too continuous, these muscles lose their tone and become weak.

All of these difficulties arising from the continuous use of incorrect shoes put a severe strain on any efforts to maintain good posture and effectively functioning back, leg, and abdominal muscles. In young children correct shoes are of greater importance than exercises in maintaining good foot position: Where correction is needed, a corrective shoe should be used along with corrective exercises.

A Correct Shoe

A correct shoe has the following qualifications:

Length

The over-all length should be such that when carrying heavy objects there is room for the great toe to spread out, since it is pressed still harder against the sole because of the extra weight. As a rule the shoe should be ½ inch longer than the length of the foot when standing on it with only the body weight to support.[7]

Width of Toe

The width of the shoe should be the same as the width of the foot at the base of the toes when standing in ordinary standing position with no weight to bear except body weight. A width greater than this will permit the foot to slide about in the shoe and this rubbing will cause corns and calluses. A width less than this will hold the toes too tightly and the pressure will cause corns and calluses.

---

[7]Shoes used for strenuous hiking should be enough longer and wider to permit the comfortable wearing of two pairs of hose, either one or both of which are woolen. This furnishes an excellent cushion and padding against stones and pebbles and allows room for the feet to spread under the long-continued body weight with the probable additional weight of a knapsack.

Vamp

The material on the top of the shoe should arch up high enough from the sole so as not to compress the toes or the instep. Such pressure interferes with toe action and causes corns on top of the toes. Also it interferes with the arches, pushing them downward thus elongating the foot and weakening the arches. Constant depression of the anterior arch shows up in the tendency to wear a hole in the sole of the shoe at that spot.

Width of Heel

The width of the heel should be the same as the width of the heel of the foot. It is difficult to understand why shoe manufacturers are so hesitant to build shoes in varying heel widths. Occasionally it is possible to find a make of shoe that does come in heel widths as well as in toe widths. They are called "combination lasts." The average shoe is too wide in the heel for many people and permits the heel to slip up and down and sideways causing blisters.

Height of Heel

Three-quarters to one and one-half inches is a proper height of heel for school and business wear. Higher heels should be worn only for dress occasions when the wearer will be on her feet but very little. In ballroom dancing the body weight is lifted up onto the toes so that the heel height is of little consequence in this activity even though the toes of such shoes are invariably incorrect and foot damaging. However, in country or square dancing there is vigorous use of the entire foot and enjoyment of the activity depends largely upon the use of low heels.

Height of Heel Box

The stiff counter in the back of the heel should not be so high that it presses against the Achilles tendon when standing on the toes. Wearing such a heel box can cause pain in that tendon.

Inner Border of Shoe

The line extending down the inner sole of the shoe should be very nearly a straight line so that the great toe will not be pushed towards the center of the foot as happens in a shoe with a curving inner border.

Shank (the part that fits up under the arch)

Dr. Frank R. Ober of Boston declares that there is no reason why shoe manufacturers cannot make shanks that can be adjusted to the individual foot and at no additional cost in the making. In most shoes the shank is made of steel and will break if adjustment is attempted.[8] Some manufacturers make the shank higher, others make it lower, and it is up to the buyer to hunt the make that fits his arch the best.

The shoe that turns its toes up after being worn a while is the shoe without a good shank.

Materials to Avoid

Composition soles cause the feet to perspire and therefore should not be used for general workday wear. Patent leather is not porous. It also causes the feet to perspire freely.

Style

The toeless shoe has its advantages since it gives good air circulation to the entire foot. Perspiring feet discourage the cure of athlete's foot, hence toeless shoes are helpful in this trouble. Many boys and men would profit by wearing such shoes. But the "heelless" shoes (the term being used to refer to shoes with no heel counter and not to shoes without heels) is another story. You need only to watch women wobbling along with feet slipping from side to side on the heel sole to realize what a strain they are under, whether conscious of it or not. Such shoes may be permissible for dress wear or other occasions where there is to be very little weight bearing on the foot, but they are a poor substitute for a business or walking shoe.

Indian moccasins and all such shoes may be good for people with strong feet when used at camp, in the home, or when walking on soft ground. They are not good for people with weak feet nor for use on concrete.

---

[8]Frank R. Ober: Shoes and Feet. J. A. M. A. 114:1554 (April 20) 1940.

When Trying on Shoes

It is important to check on correctness of over-all length; heel-to-toe length; width of toe and heel; height of vamp, heel, and heel box; inner border line; shank; and material; and then if they pass all details of this inspection it is necessary to put on both shoes and walk up and down the length of the fitting room to get the feel of the shoe in action. A person should never buy a pair of shoes without thus trying them out.

Worn Shoes

Shoes should be kept in repair, with heels leveled off when they show signs of wearing. For work hours it is important that the shoes be comfortable, therefore old dress shoes should not be "finished up" on house or school work.

Hygiene of the Feet

Pupils should be taught to wash the feet daily and for protection to use talc powder freely between the toes so the feet will be kept dry in the places most given to perspiration and hence most susceptible to athlete's foot. Care through strengthening exercises has been discussed in Chapters IV and V.

Teaching Suggestions

The teacher should plan to take some time in the fundamentals course to inform her pupils about (1) the structure of the foot, (2) its function and relation to good posture, (3) hygiene of the foot and (4) correct shoes in all details. She should have a display of various types of shoes being worn commonly at the moment, and should explain the good and poor features of each. Also there are all sorts of interesting contests which she can organize to arouse interest, such as tag days for wearers of correct shoes (date to be kept secret), contests for shoe and foot posters, and honor rolls for those who consistently wear correct shoes to school.

SMOKING

Since the school girl and boy (even those of normal college age) do not yet have their full growth too much cannot be said to them on the subject of smoking. It is absurd to work hard and faithfully at making pupils physically fit through good body mechanics and body conditioning and at the same time to offer no help in making correct decisions about bad habits that can destroy the good work of physical education.

Smoking by Girls and Women

Just how prevalent the habit is among school girls is not known for sure, but most people agree that it has increased materially among them in recent years. Up until the time of World War I a college girl who was known to smoke would have been dismissed from college. Gradually older women started to smoke and of course school girls followed their lead. As more and more women teachers, mothers, and social leaders took up the habit the more was there close at hand example that "spoke louder than words," until by now the woman who does not smoke is, in many groups, placed on the defensive.

Fortunately many of the outstanding women professional leaders do not indulge in smoking and so are in a position to advise young girls effectively on the wisdom of not cultivating this drug habit.

When smoking was finally becoming noticeably prevalent among girls in the late 1920's one writer spoke out in words that are still pertinent:

> Girls . . . make the point that they have as good a right to smoke as have men: as to this they are correct. But for neither sex is the advisability of contracting the tobacco habit a matter of legal right or equal rights. Instead, it is only a question as to whether the habit, if contracted, would add or subtract from the satisfactions of life . . . Whoever knows girls knows that, if left to themselves, they will fall victims to the cigarette propaganda even more easily than boys . . . They have certain temperamental qualities that make them particularly susceptible to shrewd advertising appeal. Almost every girl would like, if she could, to be the Duchess of Sutherland, and if she cannot be the duchess, she would like to do as the duchess does . . . Girls begin smoking to demonstrate that they are strictly modern and up-to-date in their views and habits of life.[9]

Facts About Smoking

Teachers should give their pupils scientific facts about smoking; mere opinions will not do. There are always many young

[9]Allen Benson: Smokes for Women. Good Housekeeping, August, 1929, p. 193.

people who will listen to reason and to facts and who like to have safe guides for charting their future. These pupils the teacher can guide in correct thinking if she will take the trouble to explain to them some of the researches that have been carried on in regard to smoking, giving them the results of the research and leaving it to them to arrive at their own conclusions. Mere opinions and prejudices for or against smoking have no place in these discussions.[10]

Attitudes About Smoking

This topic is one that can be pursued endlessly. It suffices here to say that in all probability the problem of smoking is looked upon by most people as a health problem and not as a moral issue as a few would rate it. But a youth's decisions about the important health habits he wills to cultivate can tellingly affect his life pattern. This applies to girls as well as to boys and it has been excellently stated in the following:

> The youth who looks forward to physical efficiency as well as mental efficiency as important factors in doing a work and achieving a place in the world should leave tobacco alone. The growing boy is injured by the use of tobacco. His growth is interfered with, his heart is made irritable, and his stomach disturbed. If the boy thinks he wants to smoke, he should wait until he is twenty-five years old; then with developed body and a wiser mind, if the use of tobacco seems desirable, let him make the choice, cognizant of its dangers and limitations. The youth who looks forward to excellence in athletics, to achievement in business or the professions, to authority and control in store and factory, will select his habits as carefully as his friends, his foods as carefully as his facts, and he will leave tobacco out of the things that are for him.[11]

Teaching Suggestions

The teacher should give the leading facts on smoking to her pupils, inviting open discussion among them so that they may clarify their own thinking on the subject. The setting of correct

---

[10]For quick information the teacher should turn to Arthur H. Steinhaus and Florence M. Grundeman: Tobacco and Health—Some Facts About Smoking. New York, 347 Madson Ave., The Association Press, 1941, (price: 35 cents). This pamphlet gives a résumé of recent scientific experiments dealing with the effects of tobacco on health. Where there is a difference of opinion among scientists on certain aspects of the subject the authors report the conflicting experiments. Any good textbook on hygiene will give the teacher material on this subject.

[11]From Dean Franklin Smiley and Adrian Gorgan Gould: A College Textbook of Hygiene, 1928, p. 172. By permission of The Macmillan Company, Publishers.

attitudes toward smoking is more important during school years for the average pupil than the dissemination of reports on scientific research. However, young folks should realize that much scientific research has been done and is being done and that the health facts are based on these findings.

An excellent leaflet on smoking for distribution to young people has been published by the National Education Association, Washington, D. C., entitled "Shall I Become A Smoker?"[12] It contains scientific information told in an attention-holding way. It tells the story simply, then leaves it up to the boy or girl to make his or her own decision. A teacher wishing such easy help in this problem should have these leaflets for distribution to all physical education classes. And it is important to reach children before they have made a decision.

WATER DRINKING

Increased muscular activity, also anxiety and fear, require increased water and salt intake for the body. The salt helps the body retain its water content. Many athletes and manual laborers realize this and take moderate amounts of water with a little salt added to it before starting strenuous exercise. At the same time the act of drinking water relieves feelings of tenseness.

As far as exercise is concerned with water drinking, it is healthful to drink lots of water, but there should be little intake just before and during the exercise, reserving the period following the exercises for the large, and then welcome, doses.

WEIGHT CONTROL

Because of poor distribution of fatty tissue many girls and women think they are overweight when it is actually bad posture, bad body mechanics or poor conditioning that is at fault; these people can be helped materially by exercises. But if there is a real surplus of fatty tissue, it is the result of either overeating or some glandular disturbance, or both, and exercise alone will not remedy it. If the cause is overeating then there is but one remedy—eat less—unless the person is willing to

[12]Written as Personal Growth Leaflet No. 103 by Arthur Steinhaus. Thousands of these have been issued. It sells for a penny or so per copy.

pay the terrific price of reducing by exercise in order to burn up the excess. The exercise requirements to burn up 1¼ pounds of body weight are staggering. No sane person would deliberately choose this punishment. For a discussion of such requirements see The Role of Food in Exercise in Chapter II.

During World War II the Federal Security Agency through the U. S. Office of Education took deep interest in the matter of correct weight of girls and women and issued the following statement:

> Weight. In order to function properly and to take one's rightful place in the busy, democratic life of our country, each woman should strive for properly proportioned weight within reasonable limits. A woman who is too heavy finds that excess weight hinders her in the satisfactory performance of her duties, and a woman who is too thin often finds herself nervous, high strung, and lacking in the high type of morale needed in our present wartime set-up. Those who are underweight also find their resistance to disease not so high as it should be. All kinds of infections find ready growth in the bodies too weak to resist their attack.
>
> Correct medical advice on the proper procedure for weight regulation is imperative, but the committee recommends that too strenuous dieting be eliminated in favor of exercise. There is too much aim for mere thinness, with consequent loss of energy and efficiency. Many women, including college girls, think of weight in terms of appearance rather than in terms of healthful physical being. There is no nutritional shortcut to fitness, no matter how effective it may be in slimming the silhouette—no vitamin pill that can take the place of exercise in building stamina and endurance.[13]

Physical education teachers should stress continuously that it is nutrition that largely determines the amount of weight while exercise influences the form of the weight, the body contours. However, this exercise responsibility is a large one and it is quite possible that whatever exercise a person performs in the hopes of reducing will be of benefit in other ways even if it proves disappointing as a reducer. But the teacher should not permit her exercise program to fly under the false colors of the reducing flag.

On the other side of the story of weight control as related to exercise is the underweight girl who is too frequently sent to the rest room for her physical education period because she is underweight when instead, unless she has a serious heart ailment, she seriously needs an exercise program to build

---

[13] Function of Physical Education in American Democracy, Part II: Adjustment of the College Curriculum to Wartime Conditions and Needs. U. S. Office of Education, F.S.A., 1942, p. 5.

strength and endurance and to aid her body in its functioning towards sound health.

Dieting Fads Among School Girls

There is a great need for all teachers to be on the alert to keep schoolgirls from the unwise dieting which many of them do in the misguided desire to keep slender. If a girl is seriously overweight dieting is probably desirable if the use of reducing drugs is not involved, but the dieting done by normal-weight and even underweight girls in the hope of acquiring a figure like that of some movie star is so fraught with danger to health that the teacher must be on the alert to check it in every instance. The physical education teacher has unusual opportunities to become informed about the dieting fads that are sweeping the school at the moment since girls exchange confidences quite openly in the "gym" dressing rooms. The alert teacher spends much time around the dressing rooms and she thus has opportunities to join in these conversations and guide the trend of thinking on many subjects. She should keep informed on the latest educational trends in the field of nutrition. The home economics teacher can keep her supplied with proper literature. Also the latest editions of the best books on personal hygiene will give her the background of information she will need in her informal, impromptu chats with girls on this subject.

Exercises to Control Weight Contours

If a girl is inclined to be stout and at the same time careless of her posture she is sure to have developed sagging breasts, flabby gluteal muscles (buttocks) and "spare tires" about her waist. Such difficulties exercise can remedy. The exercises of the conditioning series described in Chapter IV are all excellent for this purpose, particularly the Trunk Twister, Trunk Limbering, Full Squats, Sit-Ups, and Hip Roll. Other good exercises are those suggested in Chapter V, especially the exercises for the abdominal and lower back regions. Along with these exercises it is important to work diligently at correct posture. Some seeming overweight is frequently merely poor posture. Where actual overweight is present poor posture announces its presence readily while good posture tends to conceal it.

The teacher's chief concern need not be for the overweight girls. In the growing years the underweight girls are the greater health hazard, and, as a rule, they have little alarm over their condition: It is the overweight girl who gets excited about her weight. The underweight girl should do daily, general body-building exercises and engage in much outdoor sports, as much as her other physical limitations permit. The physical education teacher will need to prod her and may even have to disagree with her parents who will be fearful of exercise for her and will want her to be assigned to a "rest" class. The physician is, of course, the final judge about such matters.

TEACHING SUGGESTIONS

The teacher should impart as much hygiene information as possible in connection with the physical education classwork, especially information on topics related to exercise. As far as other hygiene topics are concerned this responsibility should no more be that of the physical education teacher than that of any other teacher in the school system. In giving these talks she should be as brief as possible, omitting all actually nonrelated material unless the pupils ask questions requiring it. Also she should talk to the point. Much of the talking can be done as the pupils are exercising. Poster displays can cover much of this material, merely requiring that the pupils' attention be called to them. However, if used, displays should be unusually well planned so they will attract and hold the students' attention long enough for the information to get across.

CHAPTER IX

TEACHING TECHNIQUES

Specific proposals for teaching the conditioning exercises and the body mechanics of everyday life have been given in the respective chapters on these subjects. The suggestions here are more general and are directed toward the teacher, motivation, class management, the art of teaching exercises in general, and safeguarding the pupil.

THE SUCCESSFUL TEACHER

To be successful a teacher should be forceful but not domineering. She should be patient, cheerful, and self controlled. She should have a sense of humor. She should be a sympathetic and understanding friend. Her personality should radiate enthusiasm and energy. She should be a keen and skillful observer. Above all she should be able to imbue her pupils with her own strength and conviction of purpose.

Guideposts

The suggestions offered below are intended to serve as guideposts for the development of the positive qualities of leadership necessary for successful teaching in physical education.

1. Be aware of the influence of suggestion by being an example in posture, appearance, neatness, and cleanliness.

2. Be friendly and show a genuine interest in each individual member of a class both inside and outside the gymnasium and allow for time during the day for informal chats. Whenever possible be dressed and on the floor ahead of the class. This offers golden opportunity for getting acquainted with pupils.

3. Be understanding and sympathetic.

4. Be genuinely enthusiastic. Classes are quick to sense the mental state of the teacher. Voice, expression, and attitude should radiate both enthusiasm and energy.

5. Make each class member feel that you are aware of her as an individual. Stimulate the laggard, encourage the timid, admonish the lazy, and hold back the pupil who hopes to work wonders overnight and as a result works too hard.

6. Be thoroughly familiar both as to theory and practice with the exercises you are to teach so that you will be able to make accurate explanations and demonstrations.

7. Memorize the order of the exercises to be taught in each lesson and if necessary all of the signals to be used. A less thorough preparation results in a frequent, frantic mental search for the next exercise and therefore less thought to the class. A teacher should be free to concentrate her whole attention upon getting satisfactory work from the group. Only on rare occasions should a teacher consult an exercise memorandum in front of her pupils.

8. The class will reflect your energy or lack of it. Therefore make all demonstrations indicate the kind of activity expected from the class—vigorous, accurate, and spirited.

9. When sensing a letdown in the activity of the class be ready and willing to participate actively in any exercise. If not done too frequently this will stimulate a class to renewed effort when other methods have failed.

10. Give every class a real workout. Most students prefer a reasonably stiff workout. A few lazy individuals may groan or complain but the majority will prefer to feel that they have put in the time to good advantage. Some teachers court popularity by giving easy work. The worth-while students respect only those teachers who offer them a challenge for hard work.

11. Motivate toward a gradually increasing individual capacity without strain. Remember that most people are inclined to stop short of their real "physiological limit" and so urge them to continue beyond the first signs of fatigue. However, do not make any child feel compelled to continue an exercise too far beyond his "psychological limit" until he has gradually built up courage and will power to urge himself on.

12. Inform your pupils about muscle soreness and lead them to expect a certain amount of it if they are to gain in strength and endurance. Undue muscle soreness, however, means that more was demanded of the muscle than it was in condition to produce at that time.

MOTIVATION

Motivation should be something more vital and important than merely oiling the machinery of a single class period so that it will run smoothly. Goals should be set up which reach far beyond the class period, the school year, or even the sum total of organized school life. Students should be made to understand that it is impossible to buy a lifetime of physical fitness by the payment of participation in vigorous physical activity up to the age of sixteen or twenty years only. They should be made to understand that the attainment and maintenance of physical fitness is a lifetime project, but that it is worth while, worth working at, and worth sacrificing for.

Constant Good Teaching and Thorough Presentation of Basic Principles

Nothing the teacher can do in motivating her pupils will interest the pupils more in the course than her constant good teaching of the course. This involves her own deep enthusiasm for the subject, her own setting of a good example as one who keeps in good physical condition and uses her body effectively, her careful presentation and skillful demonstration of the material of the course, her insistence upon hard and exact work from the pupils and her constant holding before them the facts of and understanding about the basic principles underlying correct body mechanics.

The following suggestions are presented to make a course in the fundamentals of body mechanics and conditioning interesting, stimulating, satisfying, purposeful, and useful.

1. Individuals tend to repeat a satisfying performance. Therefore to insure the use of conditioning exercises outside of school or in after-school years make their performance in school a satisfying experience. This satisfaction very rarely comes during the learning process and is experienced only after skill is acquired. This demands many repetitions of the

same movements, each gradually becoming more proficient and therefore more pleasing. Instead of becoming bored by doing the same exercises everyday for a long time students experience increasing satisfaction, which by itself is a powerful motivator.

2. Post a list of the conditioning exercises in a conspicuous place and have it available throughout the year and have the students learn it. A set of conditioning exercises which a teacher hopes will become a permanent part of an individual's daily routine should be memorized. This memory pattern should include a knowledge of the movements themselves in the correct order of execution, the names of the exercises in order, and the main purpose of each.

3. Urge the class to take home a list of the conditioning exercises and to do them every day. Suggest that they be taught to other members of the family so the family group can have fun and profit from doing them together. Be very sure, however, that each pupil understands that only those members of the family who have sound heart and lungs should participate. This is particularly true of adults and should be determined by a physician. Emphasize your interest by having the class report from time to time on the extent of their own and the family participation.

4. Explain the purpose of all special exercises given the class as well as those of the conditioning series. Suggest that the appropriate exercises and information be passed on to the sister who has dysmenorrhea or the brother who is round-shouldered.

5. Educational material becomes useful only when its use is understood by the student. Make each class hour so purposeful that when Betty's mother asks what she learned in physical education class that day, instead of saying "We ran around a lot," she will say "We learned how running will develop endurance" or, instead of "We lifted lots of heavy things," she will say "We learned how to lift heavy objects without straining our backs."

6. Whenever possible make use of the universal urges which are known to motivate children. Show how some part of the work will aid them in attaining their desired goal.

7. Use a marching song occasionally to make the practice of a good walking position more interesting and stimulating. It should not be allowed to occupy too much time, however, nor should a good walking position become associated only with those times of marching to song. If the song, or at least the words, are original with some member of the class it is even more stimulating.

8. Have a great variety of posters available for use at all times.

9. The use of music as an accompaniment for some or all exercises makes the work more stimulating for some, more interesting for others, and often secures better rhythm and less tension. Whenever music is easily available, therefore, use it for a portion of the time. It is not recommended nor advisable to use music always with the conditioning exercises. Pupils should not be imbued with the idea that music is necessary and if it is not available at home that the exercises cannot be done there. The authors recommend that one semester of the conditioning exercises done without music be alternated with one semester using music, or the first half of a semester without and the last half with music. The use of the exercises without music should always precede their use with music.

10. Use several easily administered and easily scored strength, flexibility, or endurance tests to stimulate interest in improvement and keep the scores posted on class bulletin boards.[1]

(Since competition between groups is such a strong motivating force a class may be divided into two or more permanent squads and competition between these groups may be stimulated by either averaging or totalling all the scores of the squad members. Post the name or number of the superior squad. By this means a girl is stimulated to improvement both by her own desire for betterment and by pressure from the group.)

11. When equipment is available the use of posture photographs or tracings are valuable motivators if the results are placed in the pupil's hands at the time posture is discussed and

[1]Suggestions for tests which may be used in this way are discussed in Chapter X.

each makes a record of her own body mechanics. Special exercises to use in correcting her particular weaknesses should be added to this record as they are taken up in classwork.

The experienced teacher will be able to add many other suggestions to this list.

As a conductor uses the many instruments in an orchestra to produce beautiful music, so a teacher should just as skillfully use all possible motives and incentives which will incite each pupil to intelligent self-improvement.

THE ART OF TEACHING EXERCISES

Many well-intentioned teachers give their pupils poor gymnastics work because of their lack of knowledge in the art of teaching exercises. In sports and games the objective of socializing influences can be readily attained if the teacher has an understanding of the socializing implications of the activity even though she herself may be a poor performer and may have but little knowledge of the techniques of teaching sport skills. In other words, a teacher who cannot herself play basketball well can nevertheless organize and conduct a basketball class fairly successfully as far as social objectives are concerned, provided she understands the rules of the game and the social significance of such a team sport and knows how to safeguard her pupils. True it is that the skill of the average and below average pupil will be greater if she herself is skilled in the activity and knows how to analyze the skill elements in teaching the fundamentals of the sport, but the pupil who is gifted in the type of motor skills required in the sport and who has instinctive insight into what movements are fundamental to these skills will develop into an excellent player in spite of the teacher's shortcomings. The great mass of pupils lacking inherent motor ability will never become highly skilled in a sport even with a well-skilled teacher, just as the great mass of pupils lacking inherent mental ability will never become honor students scholastically in spite of the teacher's skillful handling of her subject. The foregoing explains why so many motorly unskilled and untrained teachers do nevertheless get quite good sociological and physiological results from teaching group sports. Such teachers fail only when called upon to produce highly skilled players for teams that must win. Do not misunderstand: Skillful playing is highly desirable, but only when it can be attained without the sacrifice of good social experiences and healthful exercise.

But when it comes to teaching gymnastic exercises which should be used educationally for an altogether different set of objectives from those to be attained through sports, the picture of the teacher's needed abilities changes materially. The objectives of teaching gymnastics are discussed in Chapter IV To review briefly, they are body development, body conditioning, and the prevention or correction of postural defects. To teach gymnastic exercises so as to acquire these ends is an exact science. The exercise dosage must be understood in all its physiological implications and it must be prescribed intelligently and served up accurately if the desired results are to be obtained. Both too much and too little defeat the purposes—the acquisition of strength, flexibility, endurance, and good body mechanics.

The physiological implications of exercise dosage were discussed in Chapter II and intelligent prescribing has been taken up in various chapters, but for the accuracy which is necessary in teaching exercises effectively the following material is offered.

Formality Versus Informality

Teaching gymnastic exercises effectively is an art. It takes meticulous practice to become skillful in this sort of teaching whether it be done formally or informally. Perhaps the teacher who gets the desired results by the informal method is a greater artist than the "formalist," for successful informality in most situations is achieved only through artistry and should not be attempted too soon by the inexperienced.

There is no virtue in either formality or informality for its own sake. The teacher who finds she can do her best work by the formal method should use that method while the teacher who has no "feel" for formality would make a mistake to cultivate it. "Be yourself" is a good motto for gymnastic teaching as well as for most other situations in life.

Perhaps the chief criticism of "informality" in gymnastic teaching is that it is too easily used as a "cover" for slovenliness in teaching which is due to inadequate preparation.

When used wisely and effectively it robs gymnastics of its chief criticism born of extreme formality, namely the hard-boiled, disciplinarian type of teaching. But even with the formal method, the old militarized response to such signals as "Attention," "At ease," "Rest," and "Position" is no longer considered necessary in teaching exercises. The tendency is toward informality and the elimination of all formal terminology. The important thing is not whether the classwork is presented formally or informally but whether the students have been motivated to disciplined work and whether the classwork has been strenuous enough to produce the desired results.

The Importance of Disciplined Work

No teacher need feel that she must be on the defensive if she advocates disciplined work in physical education. True it is that many teachers loudly applaud the virtues of a physical education that is purely recreational in nature, labeling themselves "advanced thinkers" or "leaders of a new day," but the teacher who feels real concern for the proper physical development of her pupils as well as for their education for future recreational hours will not be intimidated by false philosophies. She, in the best interests of the proper physical development of her pupils, will insist upon disciplined training of the body. Such training has a spiritual as well as a physical value. Those persons who learned while young to discipline themselves through a knowledge of their needs and a willingness to sacrifice to meet these needs make up the best citizens of a democracy. It is the undisciplined who make up mobs.

This discipline need never to be of a stern, strict, severe, or formal nature. Discipline can be kindly and informal yet nonetheless discipline. In fact, there is never the slightest excuse for a teacher to be other than kindly and informal in a physical education class except in emergency cases in which some particular child or group needs another special type of leadership to solve a special problem of the moment.

All teachers should have enough discipline in every class that they have absolute control of all at a moment's notice. This is not for the purpose of making pupils sense the teacher's seeming superiority but for the best interests of the group as a whole. The difference between teacher-disciplined and student self-disciplined work is the difference between ordering pupils about or "bossing" them, and motivating them to set the task for themselves or "self-direction." In the latter situation the teacher does not command, but shows how and explains and then gives the signal (it may be formal or informal) for the students to try for themselves. This signal may mean for all to act in unison or to progress singly at their own rate and rhythm, whichever course the teacher has previously explained is desirable.

Getting the Class Started and Under Way

Since the most critical and often the most difficult part of a class period is getting started and under way, careful thought should be given to the organization of the first few minutes. Nothing excuses a teacher who allows a class to start in chaos. Upon some such signal as "Roll call" or "Fall in," the class should respond in a definitely organized and attentive way, whether it be by using previously assigned places on the floor marked by numbers, the squad method in which squad leaders report absentees, or any other method.[2] There should be an exact manner of shifting from the roll call organization to that for marching, running, or jumping rope, and from this to an open-order formation for doing exercises.

Any method of getting pupils spaced on the floor for work is good to use for opening order if the result affords observation of the whole class by the teacher, an unobstructed view of the teacher by the class, and sufficient space for each pupil to move freely. In addition, the method chosen should be one that can be quickly and efficiently executed without spending time in drilling to learn it. If classwork is started by a run around the room in single file or in a column of twos, a simple and effective method of opening order is by a column of fours, sixes, or eights down the center of the floor properly spaced and spreading out to the desired distance sideways. If rope jumping is used as the first class exercise then previously assigned places on the floor marked by spots or numbers is probably the preferable method with each girl obtaining a

[2]See Mabel Lee: The Conduct of Physical Education. New York, A. S. Barnes & Co., 1937, Chapter XIV, for detailed information on various methods of calling roll.

rope as soon she enters the gymnasium and taking her place before roll is taken, or each taking her place and the leading student in each row providing ropes for the entire row.

Where the Teacher Stations Herself in Relation to the Class

When giving directions or explanations a teacher should always be visible to the whole class and facing them, but when observing and correcting she should move about to view the class from other angles. Since aimless wandering about the room is disconcerting, all moves should be decisive. She should stand silent and still at frequent intervals, calmly, interestedly, and critically observing the class in action.

Certain of the exercises in the conditioning series (see Chapter IV) should be viewed occasionally from the side for observation of faulty postural patterns or inaccurate movements visible only from this angle. This is especially true of Full Squats, Sit-Ups, Hip Raising, Push-Ups, and Wing Lifts.

How to Develop Skilled Performance

Poise, balance, and efficiency in body movements are dependent upon skillful and effective muscular power. The exercises presented in this book are intended to aid in developing this muscular power, and if skill and efficiency are also to be evolved each exercise must be performed accurately. To aid in securing this accurate performance, any new movement must be demonstrated correctly. If a teacher is not sure of her demonstrating ability she should train a few of her most apt pupils to demonstrate for her, but she should not be content to use student demonstrators exclusively. She should practice at the exercises until she can demonstrate them.

Although demonstration is the best and most accurate way of presenting new material, a brief vivid description should accompany it. Students who are visual-minded will imitate an accurate demonstration while others need a careful description, so both should be included.

The pupil's first attempts at a new exercise must be carefully corrected to insure a progressive exactness on each succeeding performance. It is a mistake to allow incorrect execution for the sake of greater general physiological stimulation, because once a movement is performed in a certain way the tendency is to repeat it in the same way. Thus if an exercise is incorrectly performed the first time, each succeeding performance tends to be inaccurate unless a conscious, concentrated effort is made to change it. Therefore the speed with which skill is acquired will depend to a large measure upon the quickness with which adequate corrections are made of faulty movements. To allow for this, first performances of a new exercise must be in slow rhythm. In exercises where speed may ultimately be desirable it should be evolved gradually, paralleling or following the development of accuracy.

For all exercises done from a standing position the teacher should work with the class for the first few repetitions of each exercise only. This helps in establishing a correct rhythm and aids in developing unison. Then she should watch for errors in performance and make individual corrections. Whether or not the teacher works with the class at the beginning of exercises done on the floor is a matter of individual choice. Some educators believe the teacher should always remain standing except when demonstrating.

Both when exercising with the class and when demonstrating she should be careful, if facing the class, to do to her right the movements the pupils are to do to their left.

A class should be standing to observe the demonstration of an exercise to be done while standing, seated to observe one to be done while sitting down, but should never be lying down during a demonstration even of an exercise which is to be done in this position. The class must be sitting up so they can see.

How to Correct Faulty Performance

Criticism and correction must be friendly and helpful, whether admonishing the class as a whole or censoring an individual performance. It is hoped that the suggestions which follow will help a teacher not only to become more skillful in observation but also to secure the right response to criticism and correction.

1. Too many corrections given at once disperse the attention and defeat their purpose. Correct the most important faults first and give pupils a chance to produce results before adding other corrections.

2. Admonish and correct carefully to secure the desire of the class to improve.

3. At first make all corrections or criticisms general to the class as a whole. Follow this with individual criticism for the few whose performance is still inaccurate. To determine this, view the class from different angles while it works, then walk through the class and make quiet suggestions to those needing instruction, making them as inconspicuous as possible.

4. Encourage, admonish, or stimulate as the occasion demands. Be as alert and quick to praise as to criticise.

5. Commend good performance simply, briefly, and promptly in a sincerely pleased voice by such remarks as "Good!" or "That's better!" If a whole lesson has been unusually well performed, before dismissing it, inform the class of this fact and of your satisfaction.

6. Some admonitions may be in the nature of warnings to avoid certain mistakes noticeable during the previous executions of the same exercise or to avoid errors which experience has taught can be expected. These should be given before the class starts the exercise and should be as brief and to the point as possible.

7. Most corrections should be in the nature of positive hints rather than negative criticisms, such as "Tense the abdominal muscles" in preference to "Do not let the abdomen sag," "Stretch up tall" rather than "Do not slump," and "Keep the arms at shoulder level" rather than "Do not let the arms drop downward."

8. Comments designed to stimulate to increased vigor in performance or to correct slight inaccuracies may be given while the class is in motion but should always be said in the rhythm of the movements.

9. Encourage the class to breathe freely and normally. Holding the breath hinders the free circulation of the blood.

10. Always be aware of the postural pattern, constantly stimulating and encouraging toward better poise.

11. Continual observation of foot positions is also necessary. Frequent admonitions to "Keep the big toe down and roll the thighs outward" should be given. See Chapters V and VII for more information about the feet and correct foot position.

How to Attain the Purpose of an Exercise

When correct execution has been mastered, the teacher's attention should be directed toward the purpose of each exercise and toward the type of action and progression which will accomplish this purpose. In flexibility exercises this means that as soon as smooth, relaxed, flowing action has been substituted for stiff, jerky movements, emphasis and stimulation are concentrated upon increasing the range of the movements. In building endurance this means developing ability to continue strenuous exercises for a longer time with less breathlessness, less general fatigue, and less localized muscular fatigue. In developing muscular strength this means gradually "overloading" a muscle from day to day by increasing either the number of times the exercise is performed or by increasing the speed. For example, it means that if you usually do fifteen Full Squats you "overload" when you do twenty. If you usually do fifteen Push-Ups in forty-five seconds you "overload" when you do fifteen in thirty seconds. If you usually do fifteen Sit-Ups with the hands stretched toward the knees you increase the intensity when you do fifteen with the hands clasped behind the head.

Signals to Use

To start a group on unified action of a given movement, some type of signal must be given. It must include an explanation of what is to be done, followed by a brief pause after which comes the final word which indicates that action is to start. Thus a starting signal really consists of three parts, each of which will be discussed separately.

The first part should be brief and explicit, clearly indicating the exercise or movements to be done. With familiar work the name given the exercise will suffice but with less familiar work it may be necessary to follow the name with a short, concise description of the movements. As the work becomes more familiar this descriptive explanation can be dispensed with. For example, before the exercise Trunk Limbering of the conditioning series is completely familiar, the preparatory part of the signal could be "Trunk Limbering—trunk springing downward and bending sideward," followed by a pause and a final word for action. Later the signal for the same activity would become simply "Trunk Limbering—begin!"

The second part of a starting signal consists of a pause which should be just long enough to allow the class time to translate the name or the description of the exercise into the correct action in their minds. If the pause is too short there is a tendency toward movement into action as soon as the preparatory name is understood instead of waiting for the final signal. If it is too long attention wanders. Either results in slovenly, ragged action instead of a smooth, unified response.

The third and last part of a starting signal is the word which launches the class into action. The best words to use are usually verbs because verbs indicate action. Therefore the verbs "go" or "begin" are probably best to start an exercise to be repeated rhythmically, although the occasional use of the adverb "now" will add variety. The word "exercise" has been used frequently, but short and more easily accented words are probably better. When indicating a fast or vigorous movement this final signal should be given in a short, sharp, and rather explosive tone of voice. For slower exercises slightly less force should be used.

If a teacher intends to do the first few repetitions of an exercise with the class she should respond to her own starting signal just as the class does. This often helps in keeping the class in unison in exercises where this is desirable, and also aids the teacher in establishing the correct rhythm. This is particularly useful in starting exercises done in a standing position.

Suggestions of words to use in marking time or setting the rhythm for movements were included in the description of the various exercises in Chapter IV.

The best signals to use for stopping an exercise are probably " . . . and—stop!" " . . . and—halt!" or " . . . and—hold!" or sometimes just the one word "Hold!" or "Stop!" If it is desirable to go from one exercise to the next without any cessation of movement, as is suggested for the conditioning series, then a stopping signal becomes necessary only at the end of the series or perhaps occasionally during the series to allow for further demonstration or more complete correction than can be done with the class in action. When first learning exercises, it is necessary to stop the class after each one so that attention can be concentrated upon the description and demonstration of the next one. Later this may be dispensed with.

Use of the Voice

The tonal quality of a teacher's voice should be pleasing, not shrill or high pitched. If it tends to be naturally high, make a determined effort to pitch it lower. Vary the tone of voice but always keep it well modulated. Too much sameness in intonation becomes monotonous and lacks stimulation. Do not shout; a clear enunciation is better than a loud voice. Be pleasant, calm, and cheerful at all times and be sure the voice typifies these qualities.

Do not talk too much. This is hard on the teacher's voice and on the nerves of the class. Count, admonish, stimulate, or set the rhythm when necessary but do not forget there is value in the class learning to carry the rhythm unaided. When talking is advisable, be brief, direct, and brisk. The main purpose of the work is to exercise the large muscle groups of the trunk, arms, and legs of the pupils and not the vocal cords of the teacher.

When an exercise consists of movements in more than one direction, one of which should be done with more force than the others, the inflection of the voice should indicate this by emphasizing the descriptive word or count used to mark the more forceful movement. A wise choice of words also helps. For example, in the exercise Trunk Twister of the conditioning series the use of the words "Forward-fling-forward-down" is probably preferable to "Forward-back-forward-down" particularly if the "fling" is emphasized, because the word itself indicates a more forceful movement even without the vocal emphasis.

All verbal directions and descriptions must be given in a clear, decisive voice, with just enough volume to be heard by the individual in the farthest corner of the room, no more. When a class is shouted at it becomes restless and irritable; when it cannot hear it becomes listless and inattentive. A teacher should not speak until all are quiet and listening, stopping in the middle of a sentence if necessary and looking quietly at any offending pupils or making some quiet remark. She should never show annoyance, anger, or resentment—slight displeasure, perhaps, but with no animosity.

Use of Discussion

Whatever discussion is necessary to make an exercise or its purpose clear to the class should be brief but lucid. Decide ahead of time on the best method to use and plan to say as little as possible and to say that little in as short a time as possible. An activity period in physical education is not a lecture period, so a teacher must not allow herself to become wordy. A word decided on ahead of time saves nine.

If a class is in open-order formation and in the erect standing position and will continue the next bit of work from the same position, a teacher may expect undivided attention to discussion for a short period of time only, perhaps up to thirty or forty seconds. For slightly longer periods when it is advisable to keep them in open order, have them be seated on the floor, not just relaxing into any kind of attitude, but assuming some definitely good sitting position such as: (1) seated cross-legged with the hands grasping the ankles, lifting the chest against the resistive pull of the hands or (2) seated with the knees bent upward and close together, feet on the floor, hands grasping the knees, and pulling the chest up into a good sitting position or (3) seated cross-legged with the trunk held erect and hands lying idle in the lap.

For longer discussions or talks of a few minutes or more in length an informal grouping with the class seated on the floor close to the teacher lends a feeling of intimacy and importance and fosters more alert attention.

SAFEGUARDING THE PUPIL

The human body develops power and strength only through use. Individual capacity determines just how far this power building can go, but all organically and functionally sound organisms will benefit from carefully graded doses of vigorous exercise, although a defective organism may be seriously harmed. The physician should supply the information which separates individuals into these two categories.

When a teacher has positive assurance that a group of pupils can take vigorous physical exertion with benefit, they should be given strenuous activity. Too light work has been a fault of physical education in the past, particularly for girls. The average girl has been treated as though she were made of fragile glass. She is not, she is a pretty tough organism.

Arthur Steinhaus, an authority on the physiology of exercise says:

> Precluding accidents, a normal, healthy child cannot do himself permanent organic injury by physical exertion . . . There is good evidence that exercise has never permanently damaged a physiologically sound heart.[3]

Need For Protection

This point of view of the basic toughness of the human body must not be applied too literally, however, to all individuals at all times. Pupils must be protected from strains and nerve drains. Growing children especially need to be taught how to relax; they should be given sufficient exercise for proper organic development and development of physical stamina but they should not be pushed beyond their ability to take with profit.

The dosage of exercise from which each pupil can profit should be determined by a medical and physical examination which should precede assignment to classes in body mechanics and conditioning. A strenuous workout should not be given to pupils who have organic deficiencies, infections from bad teeth or tonsils, tuberculosis, rheumatic fever, chorea, scar tissue, or deformities, or who have had recent operations or any recent illness keeping them in bed for over a week. Pupils who are seriously overweight or underweight (more than 10 per cent), who are mouth breathers, or who have hernias or chronic appendicitis should not be given vigorous exercise until the condition is remedied. However, many of these cases should not be excused from physical education altogether. A teacher who is well trained in the field of corrective or restrictive exercise will be able to give these children the help which they need for their physical development. The untrained teacher should not attempt such a program.

---

[3] Arthur Steinhaus: Fitness and How We May Obtain It. J. Health & Phys. Educ., 14:427 (October) 1943.

Signs to Watch For

Even in a group which has been recommended for vigorous activity a teacher should watch for the individual who seems to tire more easily than the average when all are doing the same amount of work. Some specific signs indicating undue strain or physical distress are: unnatural pallor, extreme redness of the face, excessive perspiration, a strained facial expression, a continuous look of weariness, stumbling, or rapid and shallow breathing. The teacher should look up the medical record of these pupils and should have a conference with each one to determine why there is a lack of average physical stamina, such as history of severe illness, lack of much physical activity, no physical education in grade school or high school, or specific physical symptoms suggesting need of a careful check by a physician. Each should be given an opinion as to the best way to improve the physical condition whether this is to watch the diet more carefully, get more sleep, correct a condition of constipation, see a physician, or do more exercising at home.

If competent medical advice is not available a teacher can check the pupil's ability to recuperate after exercise as a guide to the advisability of further participation by noting the following:

1. Does she breathe much more rapidily than usual for as long as ten minutes after the exercise?
2. Does her heart beat much more rapidly than usual for as long as ten minutes after the exercise?
3. Is she still tired from the exercise after two hours of rest?
4. Is she too tired after the exercise to sleep well?
5. Is she fatigued even the next day after the exercise?

If the answer to any or all of the above questions is yes, then this pupil should not continue to participate in the activity. The teacher should have a private talk with such pupils along the lines suggested above. When necessary, the parents should be consulted and their cooperation secured in finding the cause and the remedy for the lack of physical stamina.

Specific Disabilities

Occasionally a pupil who is otherwise strong and sound and able to endure vigorous physical exertion with benefit is handicapped by some specific disability, such as a knee injury or a weak and easily sprained ankle. These pupils need to learn how to safeguard the disability against further injury, in what way the weakness may be strengthened, and how to do this without further strain.

In the material which follows a few of the more common disabilities are discussed briefly with reference to the work recommended in this book for a course in the fundamentals of body mechanics and conditioning.

Abdominal Surgery

A pupil with a history of abdominal surgery with residual pain or adhesions should be instructed to be very conservative and careful in abdominal exertion, to do a very few, if any, abdominal exercises, and to report immediately if any soreness or pain results.

Cases of recent surgery should be excused from activity or placed in special restrictive classes where very careful exercise is done only upon direct advice from the physician.

If a year or more has elapsed since the operation was performed and there is no residuum of pain, adhesions, abdominal soreness, or discomfort, the individual may safely do any and all abdominal exercises.

Athlete's Foot

Since one of the authors in an earlier publication has covered this subject quite completely it is only referred to here with no attempt at repetition.[4] The problem of athlete's foot is still with us and no teacher of physical education activities should attempt to carry on classwork without a thorough understanding of this problem which touches the work of anyone supervising shower rooms.

Back Injuries

A pupil with a history of back injury should avoid whatever movements cause pain. This will depend upon the location and type of injury sustained. Wing Lifts is probably the exer-

---

[4]Lee, op cit. The topic is discussed at length in Chapter VIII with particular reference to protection and is referred to again in other of its aspects in Chapters VI, VII, and XII.

cise which will most often need to be avoided in these cases although Trunk Limbering or Trunk Twisting may also cause pain. In most cases of this type it is wiser to suggest that the movements be done a minimum number of times, perhaps three to five, doing them very easily with no forcing of the range of action rather than to omit them entirely. A residual weakness following injuries can gradually be strengthened, but it must be done very carefully, always stopping the exercises at the first signs of pain or fatigue. If time is taken to explain to these pupils just how such weaknesses can be overcome without danger, they become keenly interested in overcoming the difficulties.

Heart Abnormalities

Some individuals with mild functional heart disorders can safely participate in moderately vigorous activity as long as they do not push themselves, stop at the first signs of fatigue, and are not allowed to do any exercise to the limit of endurance. A well-trained teacher may safely allow such an individual to be in a regular class under close supervision but the untrained teacher should either excuse such pupils or put them in a special restrictive class.

Hernia

Occasionally the physical education teacher finds in her class a pupil who has hernia. In regard to children who had a hernia in early childhood which has been cured, the American Medical Association says that it is not advisable "to limit a child in his bodily activities or athletics after the age of ten." They go on to say, "Although this child is more liable to develop a hernia than one who has not had one before, at the same time, if it should occur following athletics, one can feel secure in the knowledge that it would have come on in a few years without the athletics." And they add, "It is the general opinion among surgeons that hernias are not caused fundamentally by straining or lifting."[5]

Although there is a greater incidence of hernia among boys and men than among girls and women, still the woman phys-

[5]Editor's Replies to Letter. Hygeia, 5:322 (June) 1927.

ical education teacher does need to know that when she does encounter a case, a physician should advise about the exercise program.

Knee Injuries

Pupils who show a history of knee injuries or who complain of pain or discomfort in the knees should not be allowed to do the Squat Jump exercise of Chapter IV nor to do the Full Squats as a test of strength as described in Chapter X. These pupils can usually do a few Full Squats safely and should be instructed to try doing perhaps five very easily and their reaction should be noted. If pain or soreness results, then all deep knee bending should be omitted. If no untoward reaction is felt they should continue doing five Full Squats for several lessons, increasing only one Full Squat at a time every four or five lessons until a previously determined maximum is reached which seems safe according to the nature of the injury. When they have reached this maximum, they should continue on that level even at testing time. On this pupil's card in the space provided for a foot and leg strength test should be recorded "knee injury" and at testing time these individuals should do whatever number they have been doing in classwork and then stop.

Overweight

The overweight pupil must learn to guard against strain on the knee joint and the foot in such activities as running, rope jumping, Full Squats, and Squat Jumps.

Painful Arches

The pupil who complains of pain in the forepart of the foot (so-called ball of the foot) should omit running, rope jumping, and Squat Jumps until the arches have been strapped up or otherwise taken care of. Full Squats should be omitted if they cause pain, otherwise not.

The pupil who has very painful longitudinal arches should be in a class where special corrective exercise can be given.

Sprained Ankles

A pupil who has a sprained ankle which has healed sufficiently to allow walking should temporarily omit such exercises as Full Squats, Squat Jumps, Foot Rocker, and running or rope jumping but should be able to do all other conditioning exercises. She should omit work which involves supporting the weight on one foot at a time or bending the ankle. When the soreness disappears, she should very gradually take up all work again. In some cases of severe sprains the ankle should be reinforced by a supporting bandage during the exercise period.

Weak Wrists

A pupil who has weak wrists, previously fractured wrists, or who complains of pain in the wrists when the weight is supported on them should omit all exercises in which a good share of the weight is supported on the hands, such as in Push-Ups.

Menstrual Functioning and Exercise

Medical authorities give the following advice concerning menstrual functioning and exercise:

> There is no physiologic reason why moderate exercises should not be carried on during the menstrual period. As a matter of fact, for the average normal girl a certain amount of exercise is beneficial at this time, as exercise increases the circulation and thus tends to prevent pelvic congestion. The physiological changes which take place at menstruation are an increase in the thickness of the lining of the uterus, associated with dilatation of the blood vessels. This makes the uterus heavier than during the intermenstrual period. Because of the fact that the uterus is heavy and congested, it is usually advised that exercises involving jumping or heavy lifting be avoided at this time. This precaution is taken to prevent the possibility of the heavy uterus becoming displaced . . .
>
> For the girl who shows any abnormal symptoms, including pain, it appears safer to modify or limit physical activity. If pain is caused only by congestion, poor posture, or constipation, mild exercise without strain may be beneficial. Pain caused by structural abnormality or functional pathology may be aggravated by such exercise. Lacking specific diagnosis of the cause of pain, therefore, the conservative procedure is to limit exercise during the period. Diagnosis of the degree of pain is, of necessity, subjective, and the teacher has no accurate method for distinguishing the malingerer from the girl who is really suffering. Recognizing this, in the absence of specific medical information about the individual girl, it still seems advisable to let each girl decide for herself whether or not she is suffering sufficiently to require excuse from activity, in order to protect those girls whose difficulty may arise from definite physical disorder.
>
> With reference to the general effect of exercise on the body during menstruation: The danger of uterine displacement by exercise within the normal range of activity appears to be slight. The uterus is suspended by the broad and

round ligaments in such a way that it is supported by the bony pelvic floor, resting on the bladder, which serves as a cushion. It is securely surrounded by the other pelvic organs, which, being of essentially the same density as the uterus, serve to make independent movement of the uterus in response to the movement of the trunk almost impossible. More important would appear to be the possibility of lower back, sacro-iliac, or lumbosacral strain. There is some evidence of increased relaxation in the pelvic ligaments during the menstrual period. If strenuous exercise is performed with the joints in a mechanically disadvantageous position at this time, the likelihood of strain would be increased. Pressure on the abdominal cavity induced by excessive or extremely strong contractions of the abdominal muscles immediately before or during the menstrual period may serve to increase the quantity or duration of flow in some girls. In others the flow may be decreased or interrupted. While it is true that many girls and women can and do participate in all types of physical activity during the menstrual period without apparent harm, the state of our present knowledge concerning all aspects of menstruation is such that the best general advice to teachers would seem to be: If there is any doubt as to the effects of a given exercise, omit it during the menstrual period.[6]

The above information is offered for the teacher's guidance in her advice to girls about their exercise program during the menstrual period and in her planning of their classwork at that time. Causes of dysmenorrhea and special exercises for its relief were discussed in Chapter V.

---

[6]Editor's Reply, Exercise and Menstruation. Hygeia, 23:955 (December) 1945.

CHAPTER X

EVALUATING THE COURSE IN BODY MECHANICS AND CONDITIONING

Tests and measurements are used in physical education for two general purposes. The first is measurement for administrative purposes: to classify students for class scheduling, to group students within a class according to physical condition, or to assist in determining marks. The second purpose is measurement for determining the needs of students for motivation, and for improvement of instruction. It is assumed here that all preliminary medical and physical examinations as well as any tests used to differentiate capacities for classification and class scheduling have been given and that all individuals in this class can safely participate in vigorous activity. The present chapter deals only with the second general purpose.

The information given is not intended to help a teacher to do research in physical education. This should be carried on by that relatively small group who by interest, aptitude, and training are best suited to do research work. The purpose here is to suggest sources of more complete information, and then to describe a few of the simpler tests which an untrained or inexperienced teacher might wisely attempt.

The tests included here are by no means the only ones a physical education teacher should use to check on the effects of the total program. They are recommended to measure the results of instruction in a class in fundamentals of body mechanics and conditioning only.

WHAT TO TEST

In order to select tests wisely, a teacher should first know what to test. This in turn must be defined in terms of aims and objectives.

A single course in physical education cannot hope to provide for all desirable outcomes any more than any other single course chosen at random from the school curriculum can provide for total education. On the other hand, certain phases of education are so related that each unit is dependent upon ability in other units and may help to some extent in developing that ability. For example, the ability to read directions and figures is basic to efficiency in mathematics, and yet no teacher of mathematics would dream of using a basic test of speed and comprehension in reading as a means of evaluating improvement in mathematical ability. The same holds true for physical education. Certain elements of physical ability, such as strength, flexibility, and body control, are basic to good performance in all physical activities, but all physical activities will not develop each of these abilities to an equal degree in a given period of time. For example, although agility may be developed by jumping rope, it will not compare to the degree of agility developed through the use of games of the basketball type, where sudden starts, stops, changes of direction, and jumps into the air are almost constant. Nor is kinesthetic awareness developed by free-standing exercises to the same extent as in apparatus work or in stunts and tumbling.

Therefore the decision of what to test in a course in the fundamentals of body mechanics and conditioning is dependent upon a selection of the physical qualities which are developed most rapidly and adequately by the activities of such a course. Since these are muscular strength and endurance, relaxation, general fitness, flexibility, correct body mechanics, and a knowledge and understanding of the meaning and importance of physical fitness and how it can be attained and maintained, these are the important qualities to be measured by testing in connection with the course material of this text.

Most of the qualities can be measured, but at the present time no tests are available for use in the teaching field to measure objectively varying degrees of relaxation. Also, although many attempts have been made to devise a simple test of so-called physiological fitness based upon the heart response to exercise of a standardized amount, up to the present none of them is sufficiently accurate and economical of time to be recommended for general routine use, especially by the untrained or inexperienced teacher.

TESTS TO USE

Having decided upon what to test, the next step is to select from all tests available the very best for measuring ability in each of these categories. Whenever possible, these should always be tests which have been scientifically proven to be reliable, valid, and objective, and which are economical to use. The use of subjective measures is justifiable if they will aid in motivation and if no objective measures are available.

Muscular Strength and Endurance

There are several types of strength tests available which meet the standards of a scientific test. Dynamometer tests[1] are used to measure the utmost amount of power which a muscle group can exert in a single contraction. A second type measures ability to continue strenuous work for a short time; this is sometimes measured by the maximum number of times a standard exercise can be performed in a given length of time. A third type is used to measure the ability of a muscle or muscle group to continue work for a long time, often measured by the maximum number of times a standard exercise can be performed (no time limit set). All but one of the strength tests recommended below are of the last kind, since more girls and women are probably called upon more frequently for effort of this type than either of the other two. One is of the second type. The tests also represent the specific strengths which girls need to increase, namely, arm and shoulder girdle, abdominal, upper back, and leg and foot strength.

If the medical and physical examination has been adequate and thorough and all individuals in the class have been approved for strenuous activity, there is no danger of injuring

[1]Descriptions of some tests of this type can be found in H. Harrison Clarke: The Application of Measurement to Health and Physical Education. New York, Prentice Hall, Inc., 1945, pp. 140-145, and in M. Gladys Scott and Esther French: Better Teaching Through Testing. New York, A. S. Barnes & Co., 1945, pp. 121-122.

the heart or lungs by the prolonged performance demanded by these tests.[2]

The following tests are recommended for use as practice exercises in every instructional period, and as such have been incorporated in the suggested series of conditioning exercises described in Chapter IV. As a consequence, they require a minimum of explanation and instruction at testing time.

Arm and Shoulder Girdle Strength

The following test has been called "Floor Dips," "Knee Dips," or "Push-Ups." The term "Push-Ups" will be used here.

The position and movement are identical with the Push-Ups of the conditioning series described in Chapter IV. If the daily execution of this exercise has been correct and all cautions observed, very few explanations or instructions should be necessary before administering the test. The downward and upward movements are continued as long as it is possible to do so without a pause for rest which breaks the rhythm. An upward push must start just as soon as the chin has touched the floor (the chin must touch without thrusting the head forward to accomplish this). A downward movement must follow immediately upon a complete straightening of the elbows.

The knees may be protected by using mats or by placing folded towels under the knees if the floor is not too slippery. This is not imperative, however, and the authors have successfully used this movement, both as an exercise and as a test without knee protection. Very few girls complained, and a few folded towels were kept in convenient places around the gymnasium for the daily use of these few.

The test is performed by one-half of the class simultaneously and scored by a partner. Score the number of correct **upward** movements only which can be done without stopping. The scorer should count softly but audibly, not loud enough to disconcert others, but sufficient for the performer to hear. Any incorrect movement is not scored but if a correct performance can follow immediately without any break in the rhythm it may be scored. If the performer has done five correct Push-Ups

---

[2]Arthur H. Steinhaus, Alma M. Hawkins, Charles Giauque and Edward C. Thomas: How To Keep Fit and Like It. Chicago, Consolidated Book Publishers, Inc., 1943, p. 13.

the scorer has called softly "five." If the sixth is incorrect the scorer repeats the number of the last correct performance "five," if the seventh is also inaccurate she again repeats "five." This aids in keeping accurate account of the score and also informs the performer of an incorrect execution. When the performer begins to tire the scorer must be particularly alert to watch for a sagging back and the tendency to raise the shoulders first, then the abdomen and lastly the thighs.

If this exercise has been used as one of the conditioning series and taught in the beginning with correction by partners as suggested in Chapter IV, then each individual will not only give a more accurate performance but each will be a trained observer and scorer.

The subject of arm and shoulder girdle strength cannot be left without some comment upon the use of "chinning" exercises or some other type of "pull-up" movement as a test of arm and shoulder girdle strength. A pull-up test would be excellent but it takes equipment which is not always available, also it takes more time to administer. It is not readily adaptable to everyday exercise and when not used, how can pupils be expected to improve?

The so-called "Bookwalter Chinning" or "Straddle Pull-Ups"[3] are not very satisfactory for girls for esthetic reasons. Also girls are apt to wear their fingernails long and these cut into the palms of the girl doing the supporting, unless a wrist-lock grasp is substituted. Even though it may be used as a practice device, partner support of this kind is not steady enough nor equalized sufficiently for a testing situation. Instead, parallel bars, a horizontal bar, or a securely supported rod should be used.[4]

Foot and Leg Strength

The authors would prefer to recommend a good test of foot strength instead of one of leg strength, but no really good test of foot strength has as yet been devised. With all tests which

---

[3]C. H. McCloy: Home Calisthenics. J. Health & Phys. Educ. 14:17 (January) 1943.

[4]For detailed description of administering the pull-up test see the Committee Report: Physical Performance Levels for High School Girls. J. Health & Phys. Educ. 16:356 (June) 1945.

demand hopping on one foot there is too much danger of turning and spraining the ankle although they do have the advantage of comparing the strength of the two feet. Until a satisfactory test of foot strength is originated[5] it is probably better for the untrained or inexperienced teacher to use the test of leg strength described below. Some foot and ankle strength is also involved but not differentiated from the strength of the hip and knee extensors. Tentative standards are available and this test also has the advantage of being more easily administered since it is used as a practice exercise.

The full-squat test of leg strength is performed exactly as the Full-Squat exercise of the conditioning series which is described in Chapter IV. No particular speed is required but movement must be continuous. If the performer pauses for rest the test is considered over and succeeding squats cannot be scored. The position of the feet may be reversed at any time if it can be done without a pause or break in the established rhythm.

The test is performed by one-half of the class simultaneously and scored by a partner. Partners should count aloud softly each time the knees are **extended.** If the balance is lost and the performer touches the floor or the legs with the hands, this squat should not be scored but the count of the previous squat repeated aloud. The performer may continue if she is able to recover immediately and continue without any break in the rhythm.

The arms are thrust forward to make it easier to maintain the balance, since this is not a test of balance. Therefore, if any performer prefers to do the test with the hands kept on the hips and is able to do so without losing the balance, this may be allowed.

Abdominal Strength

The sit-up test of abdominal strength is performed exactly as the Sit-Up exercise of the conditioning series when done from the third position with the **hands clasped behind the head.** The cautions listed for the exercise in Chapter IV should be carefully observed in the testing procedure. The feet are

---

[5]The best test of foot strength which is available at present is probably the bounce test described in Scott and French: op cit.

supported by a partner. The performer continues the Sit-Ups touching right and left knees alternately as long as it is possible to maintain continuous motion without breaking the established rhythm or pausing for rest. No particular speed is required as long as movement is continuous.[6]

The test is performed by one-half of the class simultaneously and scored by a partner who also holds the feet. The score is the number of times the trunk is lifted, the partner repeating the count softly each time an elbow touches the knee. If the hands are unclasped or move away from their position behind the head, that particular movement does not count but the next Sit-Up may be scored if it is correct and follows immediately without any pause or break in the rhythm.

Upper Back Strength

Wing Lifts of the conditioning series are recommended for use in testing the strength of the upper back extensor muscles. This is not as objective as some of the other tests because it is difficult sometimes to judge the height of the lift. It is very valuable, however, as a motivator and has the advantage of being useful both as an exercise and as a test. Therefore, it is recommended in spite of its drawbacks until some better way of measuring upper back strength is evolved.

For details of correct execution and common faults of this exercise, see Chapter IV. The toes must remain in contact with the floor and the elbows must be kept up as high as possible throughout the test. The elbows must not be lowered and lifted with each movement. A folded towel may be placed under the front of the pelvis if the performer is thin and finds the pressure uncomfortable. Continue for one minute.

The score is the number of correct upward movements which can be executed in one minute. A pause for rest is allowed in this test whenever necessary if the elbows are kept off the floor. As soon as the elbows are lowered to the floor the test performance is over. Score only correct lifts, those which are high enough to conform to the set standard. Do not score any movement accompanied by a lowering and lifting of the

[6]The forward position of the head, the flattening of the chest, and the pronounced spinal flexion (curling of the spine) which are a part of the sit-up exercise are to be encouraged in this exercise.

elbows in a sort of flapping motion. The test is performed by one-half of the class simultaneously and scored by a partner. The instructor sets the time with the use of a stop watch or watch with a second hand and the following instructions: "Face down for Wing Lifts [pause], elbows up and ready [pause], go! [pause], ten seconds [pause], twenty seconds [pause], (and so on), fifty seconds [pause], stop!" The score is recorded as the number of lifts which conformed to the standard set.

If this exercise was taught in the beginning with correction by partners as suggested in Chapter IV then each individual will not only give a more accurate performance but each will be a trained observer and scorer.

Flexibility

Very few tests have been devised for measuring some of the important flexibilities, particularly those of hip and back flexion, which are simple enough to be scored by the students themselves with sufficient accuracy to be useful for purposes of motivation and measurement of improvement. If the time allotment for testing is sufficient to include a measure of flexibility, it is recommended that one of the two tests of hip and back flexion described by Scott and French[7] be used.

Body Mechanics

Many attempts have been made to devise a simple objective measure of body mechanics. Most of them have been aimed at developing a measure of anteroposterior body alignment, a few at measuring the mechanical condition of the feet. Most of them take cognizance only of standing posture. Whenever equipment, staff, and time are available, most physical education teachers include the use of standing posture tests, such as the silhouette, and of footprints as a regular part of the physical examination given at the beginning of the school year. The information obtained in this manner is very valuable and should be used in this as well as other courses as a means of discovering needs and of motivating for improvement.

It is felt, however, that in an instructional course which aims directly at good body mechanics in everyday life, a test of

[7]Scott and French: op. cit., pp. 128-131.

the mechanical use of the body in sitting, walking, and doing other common movements should be included. The one described below, although not as scientifically accurate as some measures of standing posture, is at least fairly objective in the hands of a trained teacher and has the advantage of measuring the dynamic rather than the static aspects of posture.

IOWA POSTURE TEST[8]

The Iowa Posture Test is based upon a subjective rating through observation. Experiment has proven that the results are quite correct and objective when scored by the professionally trained and experienced physical educator, less so when scored by the untrained and unskilled observer. However, it has proven to be such a valuable aid in motivating for better posture, that it is recommended here for use even by the untrained or inexperienced teacher. Although unskilled in the beginning, there is no reason why a teacher cannot improve her aptitude through study and experience. The results of each successive testing should be more accurate.

It is recommended that a three-point rating scale be used; 3 points denoting good performance; 2 points, fair performance; and 1 point, poor performance. Greater differentiation than this is too difficult and inaccurate.

The abilities to be tested and a brief summary of criteria to use in scoring each are outlined below.[9] For a more complete and detailed description of correct performance on each item see Chapter VII.

I. FOOT MECHANICS

A. Heel-Toe Walking

Rating: Good — Score 3 points

1. Heel in contact with ground first.
2. Weight transferred through to ball of foot.
3. Toes used in definite push-off action at each step.
4. Spring in walk.

---

[8]A test devised by the Women's Department of the Division of Physical Education, State University of Iowa. It includes such items as standing, walking, sitting, rising, stooping, lifting, carrying, jumping, and ascending and descending stairs. Recently they have modified the test environment from their original, more formal one recommended here to one in which several students are placed in a socio-drama situation.

[9]Adapted from C. H. McCloy: Tests and Measurements in Health and Physical Education. New York, F. S. Crofts & Co., 1939, pp. 259-263.

Rating: Fair — Score 2 points

1. Heel in contact with ground first.
2. Only slight transfer of weight to ball of foot.
3. Toes not used in push-off action.
4. General appearance of lifting whole foot with minimum of ankle and toe action.

Rating: Poor — Score 1 point

1. Toes down first.

B. Absence of Pronation

Rating: Good — Score 3 points

1. Inner line of the feet straight, no bony bulge in front of and below inner ankle bone.
2. No protrusion inward of inner prominence of ankle bone.
3. Heel cord seen in rear view is straight, not bent inward at ankle level.

Rating: Fair — Score 2 points

1. Slightly convex appearance of inner line of feet with slight bulge in front of and below inner ankle bone.
2. Slight to moderate protrusion inward of inner prominence of ankle bone.
3. Rear view of heel cord shows slight to moderate turning in at ankle level.

Rating: Poor — Score 1 point

1. Marked convexity of inner line of foot with prominent bulges.
2. Marked protrusion inward of inner prominence of ankle bone.
3. Rear view of heel cord shows marked turning inward.

C. Feet Parallel

Rating: Good — Score 3 points

1. Feet parallel.[10]
2. Toeing in is considered a strong position of the feet and as such may be rated good, but it is unattractive and should be discouraged.

Rating: Fair — Score 2 points

1. Moderate toeing out.

Rating: Poor — Score 1 point

1. Marked toeing out.

II. STANDING POSITION

A. Correct Alignment of Body Segments

Rating: Good — Score 3 points

1. Correct alignment.

Rating: Fair — Score 2 points

1. Slight general deviation or moderate deviation of one part.

Rating: Poor — Score 1 point

1. Marked general deviation.

---

[10]A very slight angle of toeing out is sometimes considered normal. This angle should probably be no more than 5 degrees.

III. WALKING

A. Correct Alignment of Body Segments

Rating:

Rate as above checking particularly on any changes from the standing position.

B. Weight Distribution

Rating: Good — Score 3 points

1. Good weight distribution, weight carried only slightly farther forward than in standing position.

Rating: Fair — Score 2 points

1. Weight carried perpendicular.

Rating: Poor — Score 1 point

1. Weight carried backward or too far forward.
   Note faulty weight distribution on chart as:
   F — meaning weight too far forward.
   B — meaning weight backward.

IV. SITTING

A. Sitting Position

Rating: Good — Score 3 points

1. Sitting position correct in all aspects.

Rating: Fair — Score 2 points

1. Slight to moderate deviation from correct standard.

Rating: Poor — Score 1 point

1. Marked deviation from correct standard.

B. Rising from Sitting Position

Rating: Good — Score 3 points

1. One foot slightly under the chair with the other a little in advance.
2. Trunk inclined slightly forward from the hips.
3. Push up in a straight diagonal line forward-upward.
4. Arms relaxed.
5. Hips kept well under body.
6. No stiffness.

Rating: Fair — Score 2 points

1. Feet in correct position.
2. Push up is slightly forward and then upward instead of a straight thrust diagonally forward-upward.
3. Arms relaxed.
4. Hips are thrust backward a bit as raised instead of being kept under body.

Rating: Poor — Score 1 point

1. Feet side by side.
2. Rising is accomplished by a marked trunk bend forward followed by a push upward.
3. Hands are used on knees or thighs to push with.

V. STOOPING TO PICK UP LIGHT OBJECT

Rating: Good — Score 3 points

1. One foot slightly ahead of other, feet and hips well under body weight.
2. Moderate bend from hips, almost complete bend from knees.
3. Relaxed and slightly rounded upper back to avoid stiffness, otherwise comparatively straight line of trunk.
4. Eyes looking downward at object.
5. Arms relaxed.
6. Smooth, balanced movement, no sudden jerks or dips.
7. Pick up object outside or only slightly ahead of foot.

Rating: Fair — Score 2 points

1. Very slight deviation from good standard in several items listed above or moderate deviation in no more than three.

Rating: Poor — Score 1 point

1. Markedly incorrect performance, especially bending from hips with straight knees.

Other items which might be included are running, jumping, lifting, carrying books or other objects, reaching, stair climbing, or any other common activity. The original of the test described above includes a sixth category, ascending and descending stairs.[11] If it is possible to add other items without too great a sacrifice of time it may be wise to do so if the activity is to be covered in class instruction.

Knowledge and Understanding

To have meaning and significance for the student, a physical education period should provide not only a good stiff workout but a certain amount of information. Any educational tool is useful to the individual only as he sees the place and use of this tool in his program of living. This implies knowledge, understanding, and appreciation as well as skill. Knowledge tests provide the means by which this understanding can be measured. They also serve as a check on the efficiency of the instructor's teaching.

Construction of Knowledge Tests

Comparatively few tests of knowledge in physical education have been published to date. More have been published in the field of health knowledge than in the various activity fields. Until such time as standardized tests of knowledge may be

[11]McCloy: op. cit., pp. 262-263.

available, a teacher must construct her own. Some teachers prefer to do this in any case.

A teacher should be thoroughly familiar with the various types of objective tests before constructing a test of her own. The present chapter makes no attempt to supply this information.[12] It includes only a brief statement of each type with a few examples.

There are two general types of objective **recall** tests which are used. In one a statement is made in the form of a complete sentence with certain important words or phrases omitted. The blank spaces are filled in by the students with the correct word or phrase. The other type makes use of a short answer in response to a key word or phrase only and the result is not a complete sentence. Neither type is recommended for use with this course.

In the **true-false** type of objective test a statement is made and the student marks it true or false. This type may be used for material where only two responses are possible. If more responses are feasible, then multiple choice would be preferable. Although the true-false type of test is widely used it is an undesirable form for physical education classwork.

Another type similar to the true-false because only two answers are possible is put in question form where the answer is either **yes** or **no.**

In the **matching** type of objective test most commonly used, two different columns of words, phrases, or definitions are given. One column may contain a name and the other its definition, both columns may have only words or names which may be paired, or one column may contain the word or phrase needed to complete a statement listed in the matching column.

The second column should contain the definitions or the statements to be completed, should always include more items than the first column, and should be numbered.

An example of the name and definition type with the answers indicated is as follows:

---

[12]Anyone interested in further study needs only to turn to the professional literature, particularly The Research Quarterly published by the American Association for Health, Physical Education, and Recreation, 1201 16th Street, N. W., Washington 6, D. C.

Directions. Match those definitions and descriptions in the right hand column with the correct name in the left hand column. Place the **number** of the definition in the space provided to the left of the name.

__6__ Muscular endurance	1. Painful menstruation
__8__ Cardiorespiratory endurance	2. A wide range of movement in a joint
__2__ Flexibility	3. Incorrect position of the foot in standing
__7__ Agility	4. Decrease of muscular tension
__4__ Relaxation	5. Exaggerated curve of lower back
__1__ Dysmenorrhea	6. The length of time a muscle can function without fatigue
__5__ Lordosis	7. Ability to move and change directions quickly
	8. Dependent upon heart, lungs, and circulatory capacity
	9. Exaggerated curve of upper back

This type of question should be used to check on knowledge of a definite type. It should not be used where interpretation, understanding, or judgment of values is necessary.

The **multiple choice** type of test seems to be the best for most physical education situations and is therefore more widely used than any other. The choice of responses to each question varies in number from three to five. There should never be less than three and there is rarely any advantage in having more than five. There is no reason to have the same number of choices for each question. The number should vary with the number of probable or credible answers. Each number should be used as the correct answer at least once throughout the test.

There are several ways of constructing multiple choice forms. One is to have an incomplete sentence or stem phrase which is correctly completed by one of the response phrases. Examples are as follows:

________1. The common cold is: (1) extremely infectious (2) moderately infectious (3) mildly infectious (4) not infectious. (Answer is 1.)

________2. When trying on shoes with the idea of purchasing, the shoes are the correct length if the great toe: (1) just touches the end without pressure (2) is ¼ inch from end of shoe (3) is ½ inch from end of shoe (4) is ¾ inch from end of shoe. (Answer is 3.)

________3. In a good standing position the knees should be: (1) bent (2) locked (3) straight but relaxed (4) together (5) apart. (Answer is 3.)

_______4. To prevent being jarred in landing from a jump one should: (1) land on whole foot and bend the hips and knees (2) land on balls of feet and let ankles, knees, and hips "give" (3) stiffen the ankles and bend hips and knees (4) land on whole foot, keeping upper trunk straight. (Answer is 2.)

Another way is to use a question form with several responses following, one of which is the correct answer. Examples are:

_______1. When lifting a heavy object off the floor with both hands what is the correct back position? (1) erect as possible (2) inclined forward (3) inclined as far forward as possible. (Answer is 1.)

_______2. A normal foot has how many arches? (1) one (2) two (3) three (4) four. (Answer is 2.)

_______3. What part of the body needs to be protected when lifting heavy objects because it is rather easily strained or injured? (1) the legs (2) the arms (3) the lower back (4) the knee joint. (Answer is 3.)

These are the two ways of constructing multiple choice forms which are best adapted to use with the informational material of a course of the kind outlined in this book.

CLASS PROCEDURE WHEN TESTING

Every detail of class organization and procedure for testing should be carefully planned in advance. This makes the testing period more interesting and meaningful for the student, more economical in the use of time and space, and makes efficient supervision easier for the teacher. There should be no opportunity for loafing and talking, thereby losing interest. If possible everyone should be kept busy all the time.

Muscular Strength and Endurance

The advance preparations for these tests may be outlined as follows:

1. Individual score cards printed or mimeographed with space for the following items: name of student, her number, her class hour or classification number, the date, names of all tests with space for several recordings and the rating on each. (See Fig. 62).

2. One card must be prepared for each girl, inserting her name, number, and other pertinent information called for.

3. Sort the cards into alphabetical order or by roll-call numbers into separate class groups.

Name________________________ Class ____________________ No. __________

Date	Push-Ups		Full Squats		Sit-Ups		Wing Lifts		Final Grade or Rat'g
	Score	Rat'g	Score	Rat'g	Score	Rat'g	Score	Rat'g	

Fig. 62. Sample individual score card for recording results of tests of strength.

4. Prepare charts showing the standards to be used for each test, including grades or ratings, and post in several conspicuous places about the room in which the tests are to be given. (See Tables II and III.)

5. Provide a box of pencils sufficient to supply one-half of the largest class.

6. Place each student's score card on her roll-call number on the floor just before class convenes.

7. Place a pencil on every even number.

Class Procedure

The class procedure can also be outlined and should follow a pattern similar to the one which follows:

1. Class run of one minute for warming up. A group or an individual should always warm up before testing to assure best performance and minimum discomfort.

2. At end of run each student picks up her own score card and remains standing on her roll-call number. Every alternate student picks up a pencil.

3. Take roll call while Number 2 above is being done.[13]

[13] If roll-call numbers are not used and roll is taken orally or by squads, the score cards may be given out by the squad leaders or by the teacher after the initial warming-up run of one minute. Roll may be taken at the same time by recording as absent all whose cards remain undistributed. It will then be necessary for the class to line up and count off in order to secure partners. The test procedure then continues as outlined.

Table II

T-SCALE SCORES FOR TESTS OF STRENGTH FOR GIRLS OF HIGH-SCHOOL AND COLLEGE AGE

T-Scale Score	0	5	10	15	20	25	30	35	40	45	50	55	60	65	70	75	80	85	90	95	100
Number of Push-Ups (1)			0	1	2	6	10	13	17	21	25	28	32	36	39	43	47	51	54	58	61
Number of Full Squats (2)			0		25		32		40		46		55		69		99		142		
Number of Sit-Ups (3)		0	1	3	7	11	15	18	22	25	29	33	36	39	43	46	50	54	57	61	65
Number of Wing Lifts (4)		29	34	38	42	47	51	55	60	64	69	73	77	82	86	91	95	99	104	108	112

(1) and (3) Physical Performance Levels for High-School Girls. Report of the Research Committee, National Section on Women's Athletics, American Association for Health, Physical Education, and Recreation. J. Health & Phys. Educ., 16:309 (June) 1945.

(2) C. H. McCloy: Home Calisthenics. J. Health & Phys. Educ., 14:18 (January) 1943.

(4) WAC Physical Fitness Rating, War Department Training Circular No. 40, June, 1944, p. 5.

Table III

STANDARDS AND GRADES FOR TESTS OF STRENGTH FOR GIRLS OF HIGH-SCHOOL AND COLLEGE AGE

Letter Grades Numerical Grades	– 60	F 65	+ 69	– 70	D 73	+ 76	– 77	C 80	+ 83	– 85	B 89	+ 91	– 93	A 95	+ 98
Number of Push-Ups		1	6	9	12	16	19	22	26	29	32	36	39	44	49
Number of Full Squats	20	25	30	35	40	45	50	55	60	75	90	105	120	135	150
Number of Sit-Ups		0	2	6	10	14	18	22	26	30	34	38	42	48	54
Number of Wing Lifts	25	32	37	42	48	54	60	65	72	77	83	90	95	100	107

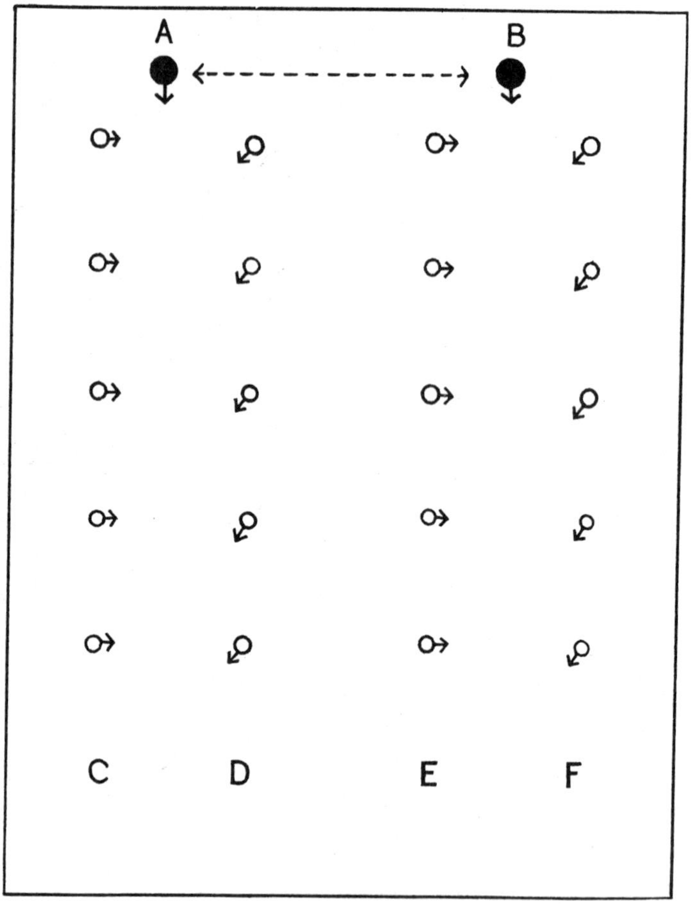

Fig. 63. Arrangement of class for tests of strength. If room is long enough one double column of paired students is preferable and easier to supervise: A and B represent the instructor; C and E, the rows of students who are scoring the test; and D and F, the rows of students who are taking the test.

4. Have partners exchange score cards, every even numbered student (2, 4, 6, etc.) exchanging cards with the student on her right (or left).

5. Have all even numbers walk four or five steps forward and then turn to face the partners whose cards they hold. (See Fig. 63).

6. The odd numbered students (1, 3, 5, etc.), who are all on the same side, will perform the test first, scored by the partners

opposite who are seated on the floor with pencils and score cards in hands.

7. Instructor gives all necessary directions for correct execution and scoring of the first test, including a demonstration of the proper technique and one or two of the most common faults.

8. Instruct students to remain in place when the test is finished.

9. Instructor then gives the following directions as to correct facing and position on the floor of those performing the test and deliberately encourages students to give an all-out effort:

a. Push-Up test—diagonal angle with head toward scoring partner.
b. Full-Squat test—diagonal angle facing toward scoring partner.
c. Sit-Up test—feet toward and held by scoring partner.
d. Wing-Lift test—diagonal angle with head toward scoring partner.

10. Instructor gives a signal for the performers to begin.

11. As each student completes the test she sits down in place and her partner records the score. Instructor notes approximate time when last performer stops.

12. When all of the first line have finished, the pencil is sent across the floor to the partner opposite.

13. Instructions are given to even numbers (2, 4, 6, etc.) as to direction of facing to perform the same test which should be diagonally toward the same end of the room but toward the opposite corner.

14. Instructor gives the signal for the performers to start.

15. When finished the performer sits down and the partner records the score.

16. When all have finished the instructor notes the time again, and if less than five minutes have elapsed since the previous checking of time the whole class is instructed to rest in place. Each student has an opportunity to rest while her partner is performing a test, and this rest period should be at least five minutes in length.

17. The same procedure from item Number 6 on is followed for each succeeding test to be performed. A period of five minutes should be allowed between test performances.

Name	Foot Mechanics			Standing	Walking		Sitting		Stooping	Total Score	Rating
	Heel to Toe Walk	No Pronation	Feet Parallel	Alignment	Alignment	Weight Distribution	Sitting Position	Rising from Sitting	Pick up Put down		
1.											
2.											
3.											
4.											
5.											
6.											
7.											
8.											
9.											

Fig. 64. Sample score sheet for test of body mechanics.

18. When all tests for that particular class period have been completed, partners exchange score cards so each has her own.

19. Collect all pencils.

20. Each girl now takes her own score card to where the test standards are posted and compares her performance with the standard. If pencils have been collected, there is no temptation to change recorded scores.

21. As each girl finishes her comparison she places her card on her roll-call number[14] and the class then proceeds as usual to the regular instruction of the period. If the testing procedure has taken the whole period, she leaves to shower and dress.

22. The cards are easily collected in alphabetical or numerical order.

When more than one test of strength is given in a single class period they should be combined into Push-Ups and Full-Squats in one period, and Sit-Ups and Wing Lifts in another. If all four are given in succession they should be in this order: Push-Ups, Full Squats, Sit-Ups, and Wing Lifts. If a test of flexibility is used it should be given first.

Body Mechanics

The advance preparations for giving the Iowa Posture Test of body mechanics may be itemized as follows:

1. Score sheets properly prepared, one for each class, with the names in alphabetical order or in order of roll-call numbers. (See Fig. 64).

2. Two small objects to be picked up in the stooping test.

3. Correct number of chairs arranged according to the diagram in Figure 65. One of the authors has used this arrangement a number of years and finds it effective. Scott and French[15] suggest a similar arrangement. Chairs may be numbered with show card paint which will easily wash off.

4. Three pencils (with erasers) sharpened and ready for use.

5. Score sheet for the correct class fastened to a clipboard or stiff cardboard before each class convenes.

---

[14]If roll-call numbers are not marked on the floor, each girl can hand her card to the instructor or the squad leaders can collect them and hand them to the instructor.

[15]Scott and French: op. cit., p. 163.

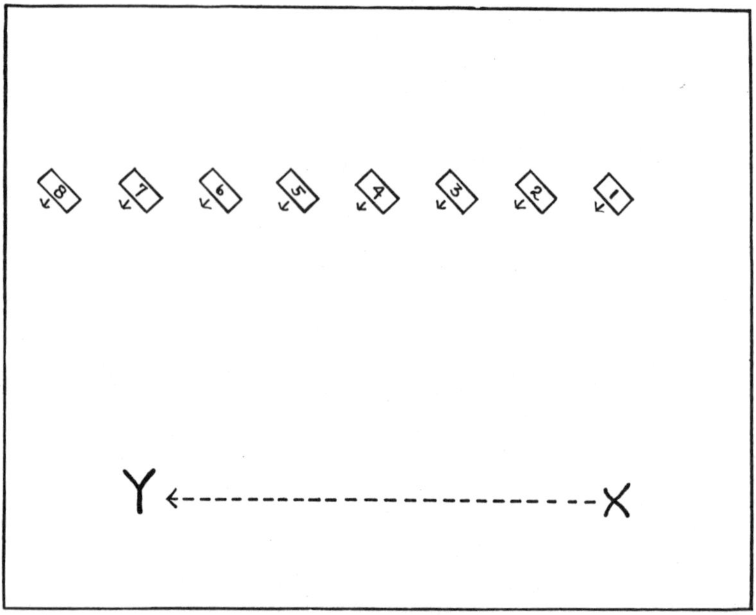

Fig. 65. Arrangement of class for test of body mechanics. The numbers represent chairs facing in the direction indicated by the arrows. The numbers on the seats correspond to roll-call numbers or numbers of the students as they are listed on the score sheet. X and Y indicate the position of the examiner during the various aspects of the test.

Class Procedure

The class procedure when giving the body-mechanics test is as follows:

1. Each student is seated in the chair numbered to correspond to her roll-call number. If roll-call numbers are not used the members of the class must be seated in alphabetical order as the names appear on the score sheet and in the order as the numbers appear in Figure 64.

2. At the start of the test the instructor stands at X as indicated on Figure 65.

3. The following instructions are given to the students: "When your name is called each one, in order, will do the following: Rise and walk diagonally across the room in the direction in which your chair is facing until you reach this line. [Indicate line or approximate place on the floor.] Turn here to face me and wait. When I look up and nod walk directly

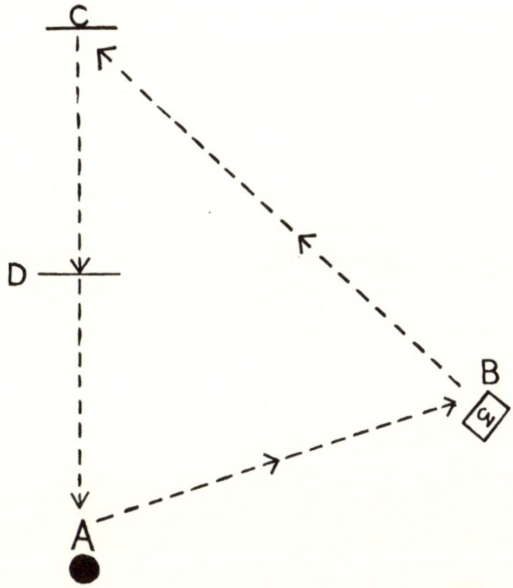

Fig. 66. Plan for rating the foot-mechanics test: A represents the instructor; B, a student (number 3) seated in a chair. As the student walks from B to C she is rated on heel to toe walking; from C to D she is rated on feet parallel; from D to A on pronation from front view; and from A to B on pronation from rear view.

toward me until 6 feet away, then turn and walk to your chair and be seated." (Fig. 66.)

4. The instructor should be able to observe and rate all three items of foot mechanics while the above is being done. Check for heel-to-toe walking on the first direction, which presents a side view, and record the rating immediately on the score sheet. Then look up and nod and as the student walks toward you check and record immediately on parallel position of feet; look up quickly to make a preliminary observation in front view on pronation before the student turns and presents a rear view of the feet as she returns to her chair. Record the rating on pronation.[16]

[16]If an inexperienced teacher finds it difficult to rate the three foot-mechanic items in succession and do it accurately, the items may be rated separately by having each student walk across the floor and back to her chair, presenting a side view while heel-to-toe walking is rated. Then next have each student walk directly toward the instructor and back again while the parallel-foot position is checked, then repeat the same while checking on pronation. This is probably the most accurate method to use for the first two or three times the body-mechanics test is given, but it takes quite a bit more time to administer. Therefore as soon as it is feasible, the first procedure is recommended.

5. As each student takes her turn, the instructor moves down the floor in the direction of X toward Y as shown in Figure 65.

6. When foot mechanics have been rated for all students the group is instructed on the next procedure as follows: "When your name is called, rise and remain standing in front of your chair looking straight ahead. When I say 'Walk' please walk across the room so I have a side view until you reach this line [indicate line or approximate distance on the floor]. Turn around and wait until I again signal 'Walk,' then return by the same path to your chair and be seated." The instructor should check standing alignment as the student stands in front of the chair, check walking alignment as she moves across the floor, and then rate weight distribution as the girl returns to her chair.[17]

7. The instructor now places two small objects on the floor about 6 to 8 feet in front of the first student.

8. The last three items can also be rated in succession for each student. Instruct the class as follows: "Remain seated until your name is called, then rise and remain standing until you hear the signal 'Walk,' then walk up to the objects on the floor, stoop and pick up one of them, rise and turn around to face in the opposite direction and stoop to pick up the second object. Then move over just far enough to replace both objects on the floor in front of the next girl. Return to your chair and be seated." Rate the sitting position of each student before her name is called. Then call her name and watch as she rises, then rate this as she remains standing. Give the signal to walk and observe the stooping position in picking up each of the two objects separately and also in replacing them on the floor for the next girl (allow both to be replaced in one stooping movement). As the girl returns to her chair, the instructor

[17]To aid the unskilled observer, the whole group may stand at once and remain standing in front of the chairs as the instructor moves down the line and rates standing alignment. Each student then moves across the floor and back as her name is called while walking alignment and weight distribution are checked. The authors prefer the first procedure because it checks more accurately upon habitual practice in good body mechanics since the student is not sure whether the instructor is rating rising, standing, foot mechanics, or body alignment. The necessary spread of attention over all of these items will result in a more nearly habitual pattern than if attention could be concentrated upon one item at a time.

records the rating for stooping. If the instructor desires to include the way in which the student is seated in the rating for rising, then observe the sitting movement and verify or change the rating which was recorded for rising from sitting.[18]

Suggestions

If the class happens to be of such a size that two rows of chairs are necessary, each chair of the second row must be placed directly behind the chair in the front row. The students are instructed to quietly and unobtrusively change places with the girl in the chair directly in front when the instructor is checking the student two or three chairs beyond. As soon as the instructor has rated the front row of chairs on one set of items, she returns to chair one and repeats the process for the rest of the class who are now seated in the front row. By the time this is done, the original girls are again in the front row and the instructor starts rating the second set of items, and so on. This procedure works very well and with no loss of time.

One of the authors has successfully rated a class of forty students in body mechanics in forty minutes by using the procedures outlined above.

Written Test of Knowledge and Understanding

The advance preparations for giving a written test are as follows:

1. Questions selected thoughtfully and carefully.

2. Questions mimeographed with space for answers on the question sheet, or separate answer sheets prepared.[19] The correct number of consecutively numbered test papers and answer sheets (if used) for each class counted out in advance and a box of pencils set ready for use.

---

[18]Here again each item may be rated separately, moving down the line and rating the sitting position of each. Then each student would rise and be immediately seated again, the instructor rating this procedure, and lastly each would walk out a few steps and pick up the objects and replace them for the next girl. As previously stated, this may be easier for the inexperienced observer to rate but is a "dead give-away" to the student on the movement being checked. The first procedure is preferable, therefore, when the instructor feels her ability will encompass it accurately.

[19]For detailed information on preparation and use of answer sheets see Scott and French: op. cit., pp. 44-47.

3. A suitable room arranged for, well ventilated and well lighted. If a classroom of sufficient size can be used (and arranged for in advance) make plans to place students in alternate seats. If the test must be given in the gymnasium, make previous arrangements for an adequate number of folding or other type of chairs to be properly spaced on the floor. In this case provide lap boards made of plywood or smooth wall board to write on and place them beside or on each chair.

4. Announce to the students at the previous class period where and when to report for the test and what to bring with them. If the test is a final examination and is to take up a whole class period, tell them to report in street clothes. If possible, all books and wraps should be left in lockers.

Class Procedure

The class procedure for a written examination which takes up a whole class period is as follows:

1. Students report in street clothes and leave wraps and books in lockers, bringing only pencil or pen to class. If lockers are not easily accessible, the wraps are placed on the back of the chair, the books are placed on the floor under or beside the chair. The student has only test questions, answer sheet if used, lap board, and pen or pencil available for use.

2. When test papers are distributed each student must receive only one copy.

3. No questions on interpretation should be answered. The directions and material should be intelligible so no clarification is necessary.

4. As each student completes the test allow her to leave; in fact, insist upon her leaving so there is no temptation to copy any of the questions.

5. As each one leaves be very sure she returns all of the test material given her, including question sheets, answer sheet, and directions. If all test papers are numbered consecutively it is simple to determine later if any test papers were not returned. If answer sheets are used the number of the copy of the test used should be placed on the answer sheet. This discourages any temptation to keep the questions because it is easy to determine who is the offender.

WHEN TO TEST

Before determining the best time to give a test, a teacher should decide what use she intends to make of the test results once they are obtained. A listing of possible uses was given in the introductory remarks at the opening of this chapter. It is not the reason in theory but the use in practice which is important. A teacher may decide that an early testing of various strengths will aid in determining class weaknesses, but unless her lesson plans from that time on are built around strengthening these weaknesses, her purpose was a theoretical one only.

With girls there is always the problem of testing in relation to the menstrual function. Recent research studies tend to show that the best scores on performance tests are made by girls immediately following the cessation of the menstrual flow; the second best scores, during the menstrual period; and the poorest scores, just preceding the onset of the flow. The variation, however, is not great enough to offset the administrative disadvantages of attempting to test each girl at the optimum time in relation to the menstrual period.

There are one or two points which must be considered in the testing program recommended in this chapter, however, in relation to menstrual function. It is obvious that the written test may be given at any time and unless the body-mechanics test (Iowa Posture Test) includes more strenuous items than are usually given it also may be given any time. The usual flexibility tests given are not strenuous enough to preclude their use at the menstrual period either. The main question of advisability lies with the tests of strength and endurance.

The authors find that the best and most authoritative information recommends that since the menstrual function is just being established and often has not yet assumed its normal rhythm for many girls of high-school age it is wiser to be conservative and omit tests as strenuous as those of strength and endurance during the menstrual period and the day or two preceding the onset of the flow. For the average girl of college age this is not necessary, except in the rare cases of individuals who should not do the Sit-Up test of abdominal strength.

The best times to give the tests recommended in this chapter will be discussed separately and in connection with a purposeful use of results.

Muscular Strength and Endurance

For purposes of self-evaluation, determining needs, or checking on the progress of an individual or a group, the greatest amount of improvement will be shown in tests if they are given at the very beginning before any practice has been done and then repeated at the end of the training period or semester. Such a procedure for tests of strength and endurance, however, is apt to result in a degree of muscular soreness and discomfort from the first testing which makes students wary on a second testing and the results are almost valueless. There must be adequate preconditioning for these tests, particularly at the opening of the school year in September when the instructor has no adequate assurance of pupils having kept themselves in condition throughout the summer. Therefore, it is wiser tc wait and test students on strength and endurance after an initial period of six to eight class hours devoted to preconditioning.

In a semester of fifteen weeks if a class meets three, four, or five times a week this first testing could be given at the end of the second week of **active** classwork. If a class meets only twice a week it is probably wiser to wait until the end of the fourth week of active classwork to give the first strength tests.

If possible, the same tests should be repeated twice more during the semester, at the end of the ninth week and at the end of the semester.

The teacher should inform pupils at the beginning of the semester that tests of strength and muscular endurance are to be used and that these are tests for which they cannot cram at the last minute (an all too prevalent educational practice) without extreme discomfort. They must understand that strength and endurance are developed gradually and it is the individual who schools herself to executing one, two, or three more repetitions of each strength or endurance exercise each class period or each week who gradually develops ability and who incidentally is able to make a good showing at test time, without the discomfort of muscle soreness. They should be instructed in the causes of muscle soreness and warned to expect it if they try to do more on the test than they have conditioned themselves for in advance.

The authors recommend that the number of strength tests to be given in a single class period follow one of the two ensuing patterns:

1. If roll call is taken by numbers marked on the floor or another method which lends itself to reorganization for the testing procedure as previously outlined in a minimum of time, the first time the strength tests are given they may be given one each day for four days at the beginning of the period preceding the regular classwork. At midsemester, two tests may be given each day for two successive class-period days. At the final testing all four tests may be given at a single class period.[20]

2. If organization for testing takes too much time, two tests may be given on each of two successive class-period days for the initial as well as the midsemester measurements. At the final testing all four should be given on the same day.

If a series of exercises for conditioning is used by an individual at home everyday, a good procedure for evaluation of progress in strength and muscular endurance is for the individual to check herself on one strength test a week. For example, she should do one test every Monday—arm and shoulder girdle the first week, leg and foot strength the second, and so on. If four tests are used, each test would be repeated once every four weeks.

Body Mechanics

If the information obtained by the body-mechanics test (Iowa Posture Test) is to be used for evaluating end results or for purposes of grading only, based upon knowledge and use at the time, taking no consideration of amount of improvement,

[20]In some series of tests it is recommended that no more than two tests of this type be given on any one day or during any one week because muscle soreness often prevents a good record being made on second- or third-day tests. This is because the tests involved duplicate the use of certain muscle groups. Such a procedure of spacing the tests recommended here is not necessary since there is no duplication of major muscular effort. Furthermore, when the four tests are given on the same day at the end of a period of training they reflect general endurance as well as the strength and endurance of certain specific muscle groups.

then the test would be administered only once, namely at the end of the semester.

If a teacher wishes to discover class weaknesses and use them as a guide to planning the lessons for the course, then the test should be given at the beginning. This also provides the individual student with information which should serve to motivate her toward improvement of individual weaknesses.

If the purpose is to check on efficiency of motivation and teaching methods, on progress of the individual and the group, for future lesson planning, or for grading where improvement as well as present status is considered, then the test should be given at the beginning and repeated at the end of the instructional course. The last is the recommended procedure.

Written Test of Knowledge and Understanding

If a teacher wishes to discover what knowledges and understandings are already known to the students and therefore do not need to be included in the plans for the course, a written test should be given at the beginning. If the test is to serve for grading purposes only then it may be given at the end. If a teacher wishes to measure the amount of individual information and understanding which has been learned, then it is necessary to give a written test both at the beginning and at the end of the semester or instructional period.

Summary

The giving of tests should not be allowed to occupy too great a proportion of the total number of instructional periods. Some testing is invaluable for motivation but too much testing becomes monotonous and uninteresting. The authors recommend the following use of the tests discussed in this chapter for a course of one semester in length:

1. Body-mechanics test (Iowa Posture Test) given the first activity class period. Time allotment should be approximately equal to one minute per student. The number of students which could be rated in thirty-five minutes by one teacher would be thirty-five.

2. On the ninth through the twelfth active class periods give one of the four strength tests each day, following an initial warm-up run but before regular class activity starts.

3. About the middle of the semester give two of the strength tests on each of two successive days.

4. If the teacher thinks a short written quiz given at mid-semester will stimulate interest and closer attention it should not occupy more than one-half of the allotted instructional time of a single class period.

5. At the end of the semester the last three or four class periods will be taken up with final examinations: one for strength tests, one for the body-mechanics test, one for the written test, and at least one for discussion and comparison of results of the final tests and inspections with the first.

INTERPRETATION OF RESULTS

The purpose of any testing procedure should be clearly understood by the pupil. If she is to gain any benefit she must really understand the test, she must know the results in terms of the scores she made, and these scores must be interpreted for her in terms of good or poor performance or improvement. She should be told these things as soon as possible after the testing procedure is concluded. If considerable time elapses between the testing and the interpretation of the results she loses interest.

The tests suggested in this chapter are not recommended as a means of predicting capacity or estimating ability. They are useful as a means of individual diagnosis, to help the teacher to devise better instructional methods, to aid the student in the development of attitudes and understandings, and as measures of attainment and improvement. They are a way of showing results, not only as a source of satisfaction to pupils but also as evidence of achievement to parents, principals, or school board. When pupils have to be graded such tests serve as a basis for evaluation.

Test Standards

If the score an individual student makes on a test is to be meaningful to her, she must have some kind of a yardstick by which she may measure her score. Such yardsticks are commonly called norms or standards of achievement, or achievement scales. Such standards should be fair and possible to achieve. For example, it would be unfair to compare the performance of young girls with standards of performance set up for young men, or the achievement of girls of nine and ten years with standards set up for girls of eighteen to twenty.

Since girls reach their peak in strength at puberty (although the exact chronological age at which this peak is attained varies), since the senior high-school age and first year of college is usually included in this period of puberty, and since a course in the fundamentals of body mechanics and conditioning as recommended in this book is designed to be given in the senior high school or first year in college, standards of performance or achievement which have been set up for either of these groups may be used here. All of the strength-test standards recommended in Tables II and III are representative of a wide group of tests. The Push-Up and Sit-Up standards are those recommended by the Research Committee of the National Section on Women's Athletics of the American Association for Health, Physical Education, and Recreation.[21] The Full-Squat test standards are those recommended by C. H. McCloy[22] and although they do not represent quite so large a group of pupils as the two tests just preceding, they were based upon test performances in various parts of the country at both high school and college levels. The Wing Lift test standards are adapted from those devised and used by the Women's Army Corps[23] and they are therefore representative of a large group of women. The age group, of course, is above the high school or early college years. Several physical educators have used these standards in freshmen college classes, however, and have found them to be satisfactory.

Some teachers prefer to construct their own achievement standards for tests, others prefer to compare results of their groups with published scales. Any teacher who wishes to construct her own scales must be thoroughly familiar with the elements of statistical procedures and must understand how such achievement scales are made and interpreted.

---

[21]Physical Performance Levels for High School Girls. J. Health & Phys. Educ. 16:309 (June) 1945.

[22]C. H. McCloy: Homo Calisthenics. J. Health & Phys. Educ. 14:18 (January) 1943.

[23]WAC Physical Fitness Rating. War Department, Washington 25, D. C., Training Circular No. 40, 10 June 1944, p. 5.

It is not necessary to have this specialized knowledge, however, in order to use norms or achievement scales constructed by others. An understanding of the meaning of such terms as "frequency distribution," "measures of central tendency," "the normal probability curve," and "correlation" is all that is necessary in order to evaluate or interpret test results.

If a teacher lacks the specialized knowledge necessary for the construction of standards of achievement it would be wiser to use achievement scales which have been made by the experts. The achievement standards given here for the strength tests are of this latter type.

If it is desirable to compare scores made on several tests or to total the scores from several tests for a composite score, the best way to do this is by using the T-scale. On a T-scale 50 always represents the mean or average. Pupils are all familiar with the term "average." If it is explained that a score above 50 is better than average (the higher, the better) and a score below 50 is poorer than average, (the lower, the poorer), then test scores made and converted to T-scale scores are meaningful to pupils.[24] Such a T-scale for the strength tests recommended in this chapter is given in Table II.

Another method of interpreting or evaluating raw scores for pupils is on a grade basis. If test results are to be used in helping to determine grades a teacher may prefer to transpose test scores into grades.

The grading standards for the body-mechanics test (Iowa Posture Test) found in Table IV are the authors' own, based upon the results of the test as given to several hundred college freshmen women over a period of several years at the University of Kansas City. If only three items of foot mechanics, one of standing, two of walking, and one each of sitting, rising, and stooping mechanics are included, this makes nine individual items which are rated from 1 to 3 points each. (See Fig. 64.) This makes the maximum possible score 27 and the minimum score 9. Table IV shows the rating or grading scale which has been used when these nine items are included. If more items are added, such as ascending and descending stairs, jumping,

---

[24]There are several sources where information may be obtained on the construction of T-scales. One of these is in Scott and French: op. cit., pp. 222-226.

Table IV

GRADING STANDARDS FOR TEST OF BODY MECHANICS

Total Score	Letter Grades		Numerical Grade
27	A+	S+	98
26	A	S	95
25	A−	S−	93
24	B+	E+	91
23	B	E	89
22	B−	E−	85
21	C+	M+	83
20	C	M	80
19	C−	M−	77
18	D+	I+	76
17	D	I	73
16	D−	I−	70
15	F+	F+	69
14	F	F	65
13	F−	F−	60

or running, then the grading scale given here would have to be adjusted.

Keeping Records

The method used to record scores on tests should be one that is simple and easy to use, can be conveniently filed away for future reference, and at the same time allows each student a knowledge of her score and how it rates. There are several ways of doing this and each may be valuable for a given test as is indicated below.

For keeping records on strength tests the individual score card is the best method. See Figure 62 for a sample card of this type with space for keeping several records on each test. These cards should be kept on file in the office by classes and each group either in alphabetical order or in order of roll-call numbers. At testing time the cards for the correct class are properly distributed, the scores recorded, the cards collected again and filed in the office until the next time they are needed.

It is easier and more economical of time to have a single score sheet for a whole group for recording the scores on the Iowa Posture Test of body mechanics rather than individual cards. Such a sample score sheet is shown in Figure 64.

If it is desirable to have in addition an individual score sheet for the student to study, a record as suggested in Figures 67 and 68 may be made from the original sheets and given to the students.

Whether individual score cards or score sheets are used to record test results they should be prepared in advance. The name of the student, the date, and all other pertinent information should be inserted before time for the test to be given. If raw scores are to be converted into T-scores or grades there should be space adjacent to the raw score for this grade or converted score.

Sometimes it is desirable to have the records for the tests of strength and endurance and the standards used for rating scores on these tests on one side of a card with space for recording scores on the other side. Such a record as this gives the student the results of several performance tests on one card.

The way of keeping a record of grades made on written tests is one of individual preference and no recommendation is made here.

Grading

When pupils have to be graded this evaluation may be based either upon actual performance and achievement or upon improvement in performance. Good arguments have been set forth at various times for both methods. Grading upon an improvement scale takes into consideration the fact that greater amounts of improvement are possible and easier to achieve at the lower end of a performance scale and harder

INDIVIDUAL RECORD SHEET

Name______________________________ Class______________ No.________

STRENGTH TESTS

Date	Push-Ups		Full Squats		Sit-Ups		Wing Lifts		Grade or Rating	Comments
	Score	Rating	Score	Rating	Score	Rating	Score	Rating		

STANDARDS AND GRADES

Letter Grades Numerical Grades	– 60	F 65	+ 69	– 70	D 73	+ 76	– 77	C 80	+ 83	– 85	B 89	+ 91	– 93	A 95	+ 98
Number of Push-Ups		1	6	9	12	16	19	22	26	29	32	36	39	44	49
Number of Full Squats	20	25	30	35	40	45	50	55	60	75	90	105	120	135	150
Number of Sit-Ups		0	2	6	10	14	18	22	26	30	34	38	42	48	54
Number of Wing Lifts	25	32	37	42	48	54	60	65	72	77	83	90	95	100	107

Fig. 67. Individual record sheet for the student's own use giving a complete record of all tests results.

Maximum score on each item: 3
Maximum total score: 27

RECORD OF BODY-MECHANICS TEST

Code: 3 = Good; 2 = Fair; 1 = Poor.

Date of Test	Score	Comment	Score	Comment
Foot Mechanics 1. Heel-toe walking				
2. Absence of pronation				
3. Feet parallel				
Standing 4. Correct Alignment				
Walking 5. Correct Alignment				
6. Weight Distribution				
Sitting 7. Sitting Position				
8. Rising from sitting				
Stooping 9. Picking up object				
Total Score				

Remarks:

Fig. 68. Rear view of individual record sheet shown in Fig. 67.

to achieve as the performance reaches the upper end. The weakness in this method, particularly when used in connection with tests of strength, lies in the fact that if pupils know they are to be graded upon improvement they fail to put forth their best efforts on early tests, thus making their scores meaningless.

The authors have found that grading in a course of the kind recommended in this book serves as a better method of motivation and ultimately results in better achievement if the evaluation is made upon the final performance level. The first test efforts thus serve as a true indication of weaknesses and if pupils are motivated and material presented properly, attention is concentrated upon strengthening these weaknesses. In the world of adult achievement we are all judged by our ability to produce results. A job usually goes to the individual who knows the most about the job to start with, not necessarily the one who will improve the most. Therefore judging work upon the basis of performance instead of improvement is one which more nearly prepares an individual for life situations.

On the basis of these arguments, then, a teacher who follows the recommendations of this chapter would evaluate a pupil's achievement along three general lines: first, knowledge and understanding as measured by a comprehensive written test; second, physical fitness in so far as it is measured by tests of strength and endurance; and third, ability to use the body efficiently in everyday tasks as measured by the body-mechanics test (Iowa Posture Test). The grades which represent achievement in these three categories may then be weighted to determine the final grade or the final evaluation of achievement in the course. The value to give to the grade in each category is suggested in Chapter XI.

CHAPTER XI

THE ORGANIZATION AND CONDUCT OF THE COURSE IN THE FUNDAMENTALS OF BODY MECHANICS AND CONDITIONING

Before the first class meeting the entire course, first as a whole and then lesson by lesson, should be carefully planned and outlined and recorded both in course-calendar form and lesson outlines. As the classwork progresses the detailed lessons can be prepared, making both calendar and lesson-plan adjustments as necessary to allow for slower- or faster-than-usual classes, and for unexpected class dismissals such as frequently occur in any school year, and for changes required to meet particular student needs and student-participation planning.

PLANNING THE COURSE

The Course as a Whole

Just how much material may be inserted in a course depends upon the length of each class period, the number of periods per week a class meets, and the number of weeks which may be devoted to the course. Also, the size of a class has some effect upon plans since the smaller the class the more individual attention each pupil may be given, and the faster the work can move along in certain of its aspects. This is a minor consideration, however, for the time element in working at the exercises is important whether the child be alone, in a small group, or in a large one.

The best plan is to prepare an outline for a course that contains all the necessary ingredients, designating in some way the minor items so that they stand out as the ones to be eliminated if, as the course develops, it appears there will not be time to include all. A good teacher does not allow lack of time to finish a course to eat into major topics.

The Statement of Objectives

Before starting work on the outline it is imperative that a statement be drawn up covering the purposes of the course; otherwise the course, if purposeful, is so only by chance. Then as each item is inserted in the outline it should be checked against the statement of objectives, and if it does not contribute to one or more of the objectives, it should not be given a place in the outline. Similarly, every main topic one can think of that does contribute to these objectives should be included to be pared down later when the outline is transferred to the actual class calendar. The objectives of a fundamentals course in body mechanics and conditioning have been discussed in Chapter III.

The Course Outline

Having determined from a study of the objectives what subject matter should be included in the course it is now necessary to (1) arrange the various topics in a logical sequence; (2) decide how many class periods should be allotted to each topic for (a) its original presentation, (b) its subsequent practice, and (c) review; (3) determine which are the minor topics which may be deleted later if necessary and (4) break down each major topic into its essential parts with their logical sequence and time allotment carefully noted.

Following this, it is necessary to list the equipment and facilities which will be needed for the successful presentation and conduct of these topics so that they may be collected together and put in order before actually needed. Also this outline should be made in advance so that if it appears that certain necessary facilities and equipment will not be available those topics requiring these things may be deleted from the outline. It is a waste of time to attempt to teach certain topics without necessary class aids at hand. The enthusiastic teacher will do all in her power to procure the necessary facilities and equipment before abandoning plans to present important class material.

The Placement of Testing Within the Course

Before the course outline can be completed and the calendar set up it is necessary to decide what tests should be given, how frequently, and where inserted in the course. The "what" decisions should be written into the outline and the "where" and "when" decisions should be written into the calendar. Suggestions as to placement are given in the sample outline and calendar which follow later.

The Policy on Class Marks Given a Place in the Course Outline

As the course is being planned decisions should be made about the various elements which will determine marks for the course so that the teacher will know from the very start what records she must keep for each pupil in order to arrive at her final marks. These decisions should be written into the course outline and should be made known to all pupils at the beginning of the course. The subject of marks will be discussed more fully later in this chapter.

The Class-Hour Pattern

The class-hour pattern depends, of course, upon the length of the class period. Once that figure is known the pattern should be determined and written into the course outline. As the course develops, lesson by lesson, this pattern should be followed meticulously except for changes occasionally required because of emergency situations or class needs. If the pattern is adhered to conscientiously the subject matter of the course will gradually unfold and be presented in its proper time with due consideration given to all parts of the course. A sample class-hour pattern is included in the sample course outline given below.

Time Allowed for Dressing

When within a class period any one of the divisions of that day's classwork uses its maximum time allotment, all other divisions of work planned for that period will have to be held down to their minimum. Five minutes at the beginning of the hour and ten at the close for dressing may seem too little but it is even more than sufficient if the students can be taught to attend strictly to business. Too many children (girls in particular) are permitted to acquire the bad habit of dawdling while dressing at home. Education should do what it can to correct this, and since it is the physical education department that must constantly supervise children in dressing and undressing situations, this educational opportunity becomes the responsibility of the physical education teacher. She should be fully aware of this responsibility and should frequently be in the dressingrooms mingling with the pupils and seizing every opportunity to teach promptness in this part of the class activity. The teacher who is firm (not cross, but firm in a kindly, joking way) about this from the very first class period will soon get cooperation from the pupils.

Too many teachers allow the pupils themselves to set the time standards for dressing, calling roll whenever most of them have at last appeared on the floor, thus wasting much time. Such a teacher does not value her class time. If she doesn't value it, who will?

An Exact Time Set for Roll Call

At the opening class meeting the pupils should be informed of the exact minute when roll is to be called. The teacher should have her watch adjusted to the official school clock and then should guard her class time jealously: The minute that moment arrives she should start the roll call, not before, not after. Pupils will see at once that she means business and they will respond favorably.

A Sample Outline for a Course in the Fundamentals of Body Mechanics and Conditioning

I. Objectives of the Course

A. To give an appreciation of the relation of physical fitness to life's activities.

B. To teach the basic principles of correct body mechanics.

C. To teach the proper execution of a set of scientifically sound conditioning exercises which can be used in maintaining physical fitness in so far as exercise contributes to total fitness.

D. To teach the value of good posture and how to attain it.

E. To give knowledge of and practice in correct body mechanics in relation to the activities of everyday life.

II. Discussion Topics

A. The objectives and content of the course.
B. Student relation to the department.
C. The place of physical education in education.
D. The place of a fundamentals course in body mechanics and conditioning in a total physical education program.
E. The need for physical fitness.
F. The basic principles underlying correct body mechanics.
G. The place of conditioning exercises in physical fitness.
H. The testing program.
I. The findings from the posture and body-mechanics tests.
J. The use of specific exercises to correct specific faults.
K. Weight control in relation to exercise.
L. Hygiene of the menstrual period and exercise in relation to it.
M. The place of relaxation in a fitness program.
N. The effective use of the body in homemaking.
O. The effective use of the body in industry.
P. The effective use of the body for charm, poise, and efficiency.

III. Testing

A. Tests given near the beginning and again at the end of the course.
   1. Physical inspection
      a. Subjective analysis by instructor
      b. Silhouetteograph or some such device if possible
   2. Body-mechanics tests
      a. Foot mechanics
      b. Standing mechanics
      c. Walking mechanics
      d. Running mechanics
      e. Jumping mechanics
      f. Climbing mechanics
      g. Sitting mechanics
      h. Stooping mechanics
      i. Lifting mechanics
      j. Carrying mechanics
      k. Pushing mechanics
   3. Tests of physical fitness in relation to strength and endurance.
      a. Push-Ups
      b. Full Squats
      c. Sit-Ups
      d. Wing Lifts

B. Written knowledge test on entire course.

IV. Practice

A. Conditioning Exercises.
B. Endurance Exercises.
   1. Running
   2. Rope jumping
C. Exercises for the correction and prevention of faulty posture.
   1. Poor alignment of body segments
   2. Forward shoulders
   3. Flat chest
   4. Prominent abdomen
   5. Incorrect pelvic tilt
   6. Weak back muscles
   7. Weak feet
D. Exercises for correction of faulty functional conditions.
   1. Dysmenorrhea
   2. Constipation

E. Body mechanics as related to everyday life
   1. Standing
   2. Walking
   3. Climbing
      a. Stairs
      b. Step ladders
   4. Running and jumping
   5. Sitting
      a. With ordinary height seat
      b. With lower-than-usual height seat
      c. On the floor
   6. Stooping, lifting, and carrying
      a. Small and large objects
      b. Light and heavy objects
   7. Reaching and pushing
      a. Small and large objects
      b. Light and heavy objects
   8. Relaxation

V. Marks or Grading

	Method I	Method II
Knowledge test	20%	33⅓%
Body-mechanics test (including posture)	40%	33⅓%
Strength tests	40%	33⅓%
Total	100%	100%

VI. Pattern of Class Period[1]

Sample Fifty-Minute Period

1. Dress .... 5 min.
2. Class .... 35 min.
   a. Roll call
   b. Endurance exercises
   c. Conditioning exercises
   d. Review exercises of two lessons ago
   e. Review exercises of last lesson
   f. Presentation of topic of day
   g. Practice on topic of day
3. Shower and dress .... 10 min.

Total .... 50 min.

[1]The United States Office of Education recommends eight to twelve minutes of conditioning exercises in eqch class period for high school physical education classes. (Physical Fitness Through Physical Education for the Victory Corps.) This refers to a class whose major purpose is not conditioning but swimming, folk dancing, and the like.

In a class where the main purpose is conditioning and effective use of the body in movement a greater proportion of time should be spent on exercises or movements for the development of strength and endurance, flexibility, coordination, and agility. Therefore a total of nineteen to twenty minutes is recommended for the development of these specifics in the time allotment given below to conditioning exerciss.

Time Allotment for the Course

The United States Office of Education recommends the following for all high schools:

1. Five periods per week of physical education class instructional work of a vigorous type.

2. Ten hours per week of athletics, long hikes,[2] school journeys, "road work" (alternate walking and running), cross-country runs, bike trips, and the like.

3. Medical examinations for all.

4. A health education program for all.

Within this time allotment devoted to physical education instructional work there should be set aside in the program a minimum of three hours per week for one semester for a course in the fundamentals of body mechanics and conditioning. In the total physical education program this course should precede all other courses, even those in fundamentals of rhythms and sports.

The teacher who needs guidance in planning these other fundamentals courses will find material on rhythms and sports by scanning the professional literature and the catalogues of offerings by publishers who specialize in physical education books. These publishers can be located through their advertisements in the official magazine of the profession, The Journal of Health and Physical Education, 1201 16th St., N. W., Washington, D. C.

The Physical Education Class-Time Allotment Combined with Health Education

In some schools the physical education time allotment must carry the responsibility for certain health education courses. In the five-hour-a-week arrangement the fundamentals of body mechanics and conditioning should use three hours per week, leaving two hours for health education. If physical education classes meet only three hours a week, the teacher should refuse to permit health education classes to steal a part of that time. Justice can be done to neither physical education nor health education when they must divide so limited a time between

[2]A hike of real physical value is one taken at the rate of at least four miles per hour on the level (less if ascending) and for a long distance.

them. The school that desires a physical education program should provide time in the schedule for that program and the school that also wants a health education progam should also provide special time for it. Every physical education hour should be a practice hour in the exercise aspect of healthful living: It should not be sacrificed in order to provide hygiene instruction which may be wanted (and is most sure to be seriously needed) but apparently not wanted enough to be given a place of its own in the school curriculum.

The Course Calendar

After the outline is made and the amount of time that can be devoted to the course determined, the subject matter should be broken down into class-period units, and these should be checked against each other and against the course outline to be sure that all subject matter is included, that each topic is given its correct relative amount of time, and that it is placed in its proper sequence in relation to the other topics. A sample course calendar is shown in Table V in Chapter XII.

FINAL EXAMINATIONS AND CLASS MARKS

Physical education classwork should close at the end of the term with a grand finale, not just "peter out" as if it had been nothing of importance. How wasteful of time classwork must seem to pupils when no check is made to see what they have learned! At the close of the course should come reviews as in other classwork. The finals should include both practical and written work covering the material of the course.

The Reviews

Students derive much fun and at the same time get a good "refresher" from the impromptu stunt-program type of review. The last half of three class periods should be devoted to this review, the first half being taken up with the conditioning exercise series. Having decided beforehand upon the main topics which are to be reviewed, the teacher writes these topics on slips of paper to "toss into a hat." She divides the class into three groups and has each group select a leader. Each leader then draws a topic from the hat and retires with her group to a corner of the room where they use the time allotted them to plan how they will present their topic to the rest of the class, attempting to bring in everything which they can remember that the teacher has presented during the course on that topic.

These practice periods are busy ones with three groups scattered about the floor, each intent on organizing and practicing its skit. For the rest of the period, each group in turn is given time to present its topic, with some time left for discussion, questions, and criticisms from the audience (the teacher and the other groups). On the second and third days the class is divided into new groupings with new leaders, and a new set of topics is drawn from the hat.[3]

A few of the recent review skits as presented by the freshmen at the University of Nebraska have been as follows:

1. A group of girls is talking together on a campus corner when a girl with poor posture passes by. They begin to criticize her. She overhears and asks, "Are you girls talking about me?" "Yes," speaks up one of them, "Your posture is so bad." "Well, what is wrong with my posture?" she asks. Then they all tell her, each pointing out a different fault. The girl takes it in surprisingly good spirit and inquires, "What can I do about it?" and they proceed to tell her that also, demonstrating their advice to her.

2. A "radio announcer" introduces one "guest" at a time to talk on some feature of a topic.

3. A "teacher" conducts a class presenting her topic. Her "pupils" ask a lot of questions until she has demonstrated and explained her topic quite thoroughly.

4. A "Town Hall" group presents a topic with a few "hecklers" challenging their statements.

5. A "lecturer" presents a lecture-demonstration bringing her "troupe" with her.

---

[3]This type of review has been used most successfully for a number of years at the University of Nebraska, using three groups each day with a preliminary seven-minute practice period and three minutes for each group presentation, followed by three minutes for discussion and criticism. It is an inspiration to see and hear the skits which the girls prepare on the spur of the moment, so creatively and cleverly bringing in the subject matter presented earlier in the course. Perhaps the greatest inspiration comes from the surprised comments of the girls themselves, such as: "I never dreamed we learned so much in this work!"; "I would never have believed that there was so much to learn in a gym class!" and "Why hasn't someone told us all these things long ago?"

Each year house mothers report to the teachers that these review periods are the main topic of conversation in the dining rooms of sorority houses and residence halls during the review days and that the whole tone of the conversations is one of pleasure over the surprising amount of interesting things they have learned "at gym."

6. A group offers a "Do This" and "Don't Do That" show.
7. A group presents a "talkie movie" with the announcer explaining the motions the actors are going through.

Before preparing their skits the groups should be told that they should bring out information on their topic as follows: (1) difficulties that may be involved, (2) how to avoid these difficulties, (3) exercises to improve the situation, (4) cautions in regard to the exercises, (5) demonstration of exercises pertinent to the topic and (6) basic principles involved.

A note of warning needs to be sounded for these review class periods. The class should be given its regular conditioning exercises. They will have to be brief to allow sufficient time for the reviewing but they can be stimulating and beneficial even though brief. No physical education class should ever leave the playing field or gymnasium floor without having had a real physical workout during the class period unless the class meeting has been of such a nature that it was necessary to request the pupils to report in street clothes.

Finals

Pupils should be given final tests in their body-mechanics skill and also should be given a final strength test with an opportunity given to compare results of their final tests with earlier tests. Also they should be given a final written examination covering the informational parts of the course. Procedures for giving tests and sample written examination questions have been given in Chapter X.

Class Grades

The grading of class achievement has been discussed in Chapter X. In the sample outline of the course given earlier in this chapter a scale was presented for evaluating the various categories of classwork for a final mark.

Many teachers take attendance records as well as class achievement into consideration in determining grades. This practice is not justifiable.[4]

[4]For a lengthy discussion of the educational philosophy back of this remark as well as for a discussion of the entire problem of marks and grading see Mabel Lee: The Conduct of Physical Education. New York, A. S. Barnes & Co., 1937, Chapter XV.

PHYSICAL INSPECTIONS

Since the course in the fundamentals of body mechanics and conditioning is only a part of the total physical education program, it is assumed that the physical education department has conducted a standard physical examination which supplements the medical examination given by the physician, and that with this combined data at hand the pupils have been classified and only those who are physically equal to this work have been assigned to take it.

Pupils should be excused from this work but rarely. A pupil who is able to attend school at all is able to profit from some form of physical education work for the improvement of her physical condition and for her education in such recreational activities as she may safely take part in. Many weak children need to acquire physical strength, stamina, and endurance if they are ever to be able to take a worth-while place in life, and they do not acquire these things except by daily vigorous exercise. By "vigorous" for some children is meant not that which would be vigorous for all children but that which is vigorous for them. As a child's level of strength and endurance is raised, "vigorous" changes its meaning for her. Too many fearful or indulgent parents and family physicians hinder children from developing properly within their physical limitations.[5]

The teacher who is untrained in the health appraisal of children should study James F. Rogers: What Every Teacher Should Know About the Physical Condition of Her Pupils, Pamphlet No. 68, revised, 1945. Copies of this pamphlet may be procured from the Superintendent of Documents, Washington, D. C. for 10c each.

Once the pupils have been classified for classwork by the physician there falls on the teacher the responsibility of inspecting the pupil's posture and body mechanics. This inspection can be given to as many as thirty-five pupils in thirty-five minutes (see Chapter X). If a special test of standing posture is given this inspection will take several class periods,

[5]For a full discussion of the physical education teacher's responsibility towards physical and medical examinations, the facilities, equipment, and supplies needed, the organization and conduct of the examinations, and the follow-up work necessary see Lee: op. cit., Chapters V, VI, VIII, and IX.

the number depending upon the size of the class and the number of other teachers available to assist with this work. At such times student leaders may be needed to take charge of those who are not being inspected in any one period.

CLASS EQUIPMENT AND SUPPLEMENTARY MATERIALS

Books, Magazines, and Pamphlets

The teacher should constantly be on the lookout for new material on the subject matter of the course. Books, pamphlets, and periodicals that contain rich source material are constantly coming off the press. Medical literature as reported for laymen is rich in scientific material for this use, as is also literature of physiological research. Our own professional literature offers much help. Certainly every teacher of physical education keeps in touch with it!

Charts

There are many charts available for the asking which make classwork more interesting and save the teacher's time in explanations. Announcements and advertisements about many of these are carried in our professional magazines. Anatomical charts in particular offer helpful material.

Films

Research on the educational value of films shows that their use is a material aid. In the field of learning motor skills (physical activity skills) it has been clearly demonstrated that the use of motion pictures is of definite value when used as a part of the regular instructional routine. Teachers should keep in touch with professional literature for announcements of current films available.

Manikins and Living Models

If at all possible a female head-and-torso manikin should be available for demonstration to girls' classes when the topic of exercises for dysmenorrhea is under consideration. Perhaps

---

[6]Jeanne Aileene Lockhart: The Value of the Motion Picture as an Instructional Device in Learning a Motor Skill; Summaries of Doctoral Dissertations. Madison, Wisconsin, University of Wisconsin Press, 1942. Vol. 7, 1941-42, p. 249.

the hygiene and biology departments could share with the physical education department in the ownership of this important piece of equipment. Carefully selected members of the class may well serve as living models. Bathing suits and leotards make ideal costumes for these models.

Miscellaneous Equipment

Chairs	Poles for vertical line tests
Jumping ropes	Posture boards
Mirrors	Sandbags
Miscellaneous light and heavy objects (large assortment)	Score sheets
	Stop watch
Music box and records	Suitcases or equivalent

There should be enough of the above objects available to equip one-half of a class at a time. If possible, there should be enough jumping ropes so that all may practice at this endurance-building exercise at the same time.

The posture boards are used to carry on the head when working at correct walking. They should be approximately 9½ x 6½ x 1 inch in size. If not available, books may be substituted but such use is rather hard on the books. It will cost very little to have the Manual Arts Department make enough posture boards for half of the largest class to use at a time.

Physical Inspection Record Card

A card 5 x 8 inches in size should be available for each student. On this is recorded the results of the posture and body-mechanics examinations. It should be so designed that it can carry the records of subsequent examinations as well as the initial one. These cards should be shown to the students after each recording and discussed with them. This can be done effectively in group conferences, a method which will save much time for the teacher.

Posters

A physical education teacher cannot have too many posters for class display. They need not be expensive. Student campaigns to build up a good collection of home-made posters are excellent class motivators. There is constantly much attractive and effective poster material available free of charge from a large number of organizations, both educational and commercial, such as the American Posture League, Metropolitan Life Insurance Company, and The National Dairy Council. A constant check on professional periodicals will keep one informed about these offerings.

It would be well to make a large poster listing the conditioning exercises used in class. An attractive framing for the list could be made of stick figures depicted doing the various exercises with one side showing correct and the other incorrect positions. Such a large poster would be very helpful not only as a motivator but as a teaching device.

Poster Making

Excellent advice for making posters is presented in the Publicity Handbook of the National Congress of Parents and Teachers as follows:

Simplicity is the keynote of effectiveness in posters.
Color in paper, illustrations, or lettering will add interest.

1. Size. Depends upon use for which poster is intended. Observe good proportions, such as 10 x 15, 12 x 16, 15 x 20, 16 x 24 inches. Commercial cardboard comes in sheets 20 x 30, 22 x 28, or 30 x 40 inches, which when cut in halves or in quarters gives a well-proportioned poster.

2. Material. Choose cardboard of sufficient stiffness that poster will not curl or bend. If cardboard cannot be found of the right color, use cover paper mounted on plain cardboard.

3. Subject Matter. A poster makes an announcement or tells a story in brief. Long explanations should be avoided. A short slogan or text is more effective than a wordy one. Use a maximum of fifteen or sixteen words.

4. Color. Bright, harmonious colors are desirable. Pastel shades are weak. Light and dark green, black and white, orange and blue, orange and black are good color combinations. For legibility there should be striking contrast between the explanatory matter and the background.

5. Lettering. Distinct from the background of the poster; should be large enough to stand out well. Both small and capital letters—rather than all capitals—should be used, since the combination is more easily readable.

Gummed letters may be secured from a stationery store, but are made in limited colors and sizes. Stencil sets or letter patterns for tracing and cutting may be used. Letters may be drawn much more quickly than they can be cut out and pasted. Sketch in letters with single pencil strokes and draw over these lines with a shading or broad lettering pen. The lettering should agree in color and proportion with the picture, if a picture is used.

6. Decoration. Helps the poster catch the eye. Striking color tones and contrasts demand attention. Soft, misty effects are ineffectual. Simple lines and gay, harmonious colors should predominate.

7. Use of Pictures. Pictures of children, women, flowers, birds, and animals have a prominent place in the poster-maker's world. Choose pictures that are done in bright flat colors. Small detail and realistic shading do not show up well on a poster. Use one good-sized picture rather than many small ones. "Let the picture tell the story."

8. General Directions. Plan the poster on scratch paper first. Allow generous margins. Avoid ornateness. Vary size of letters according to importance of words. Use simplicity of thought and wording.

The message of the poster should be simple and easily grasped. Everything about the poster should emphasize this one simple idea.[7]

Poster Displays

A large bulletinboard should be placed where students tend to gather while waiting for class to start. Several posters all emphasizing some one particular point should be put up for at least a week then taken down and others substituted. The material of the illustrations should parallel the topics of the special talks for that week or the uses of the special exercises, as for example, when discussing feet and shoes have illustrations of good and poor foot positions, imprints of normal feet and flat feet, pictures of good shoes and poor shoes, and include pictures of the cramped toes, corns, or bunions resulting from the use of ill-fitting shoes; and when discussing poise and posture show illustrations of two girls wearing the same evening dress, one well poised and the other a "sad sack" and contrast them in bathing suit, street dress, and sport's costume. Figure 69 illustrates a type of poster that was used effectively during World War II at WAC Training Centers to arouse interest in posture. The idea conveyed by this poster might be used tellingly in a poster brought up to date by using a drawing of today's schoolgirl with today's length of skirt and today's hairdress. The captions might well be changed to read:

"This Gal Committed a Postural Blunder Because Her Hips Were Not Tucked Under."

"The Day's Work Is In the Bag If Your Shoulders Do Not Sag."

"You Will Be the Pride of Your Group If You Stand Erect and Do Not Stoop."

---

[7]Reproduced by permission of the National Congress of Parents and Teachers.

Fig. 69. (Courtesy of the Department of the Army.)

Reproductions of Works of Art Depicting Good Posture

To familiarize pupils with good posture as portrayed in art, the teacher should arrange for an art display and should point out the salient features of each painting or sculptoring bearing on good posture. A suggested list of excellent reproductions to use for such a display is as follows:

Paintings

1. Botticelli, Figure of La Flora in Allegory of Spring (Academy, Florence)
2. Bouguereau, Cupid Captive

3. Breton, Song of the Lark (Art Museum, Chicago)
4. Gainsborough, The Morning Walk, Squire Hallett and His Wife (Rothschild Collection)
5. LeBrun, Girl With Muff (Louvre, Paris)
6. Leighton, "Hit"
7. Murillo, Madonna and Child (Pitti Palace, Florence)
8. Nattier, Madame Adelaide (Louvre, Paris)
9. Peale, Washington (Metropolitan Museum of Art, New York)
10. Richter, Queen Louise
11. Thayer, Caritos (Art Museum, Boston)
12. Van Dyke, William II, Prince of Nassau

Sculpture

1. Apollo Belvedere (Vatican, Rome)
2. Bartlet, Lafayette (Paris)
3. Donatello, Saint George (Uffizi Gallery, Florence)
4. Hyatt, Joan of Arc (Riverside Drive, New York City)
5. McKenzie, The Home Coming (Cambridge, England)
6. McKenzie, The Call (Edinburgh)
7. MacMonnies, Nathan Hale (New York)
8. Pomona (Uffizi Gallery, Florence). See Frontispiece.
9. Saint-Gaudens, The Shaw Memorial (Boston)
10. Ward, George Washington (Treasury Building, New York City)

Score Cards

There should be kept on file an individual score card for each pupil, on which should be recorded the results of all strength tests. For description and use of these cards see Chapter X.

Silhouetteograph or Other Pictorial Posture Record

If thorough work is to be done in posture training some device should be employed whereby pictorial comparative records may be preserved of the student's standing posture both at the beginning and at the end of the course. If time permits, a record at the middle of the course also is highly desirable in order to check on class progress before the course is ended.

The day may come when motion pictures will be cheap enough for common use in the classroom. Probably this is the ideal form of record of a student's posture and body mechanics, but for the present its cost is prohibitive. Some schools (the few) own a camera-device and can afford actual negatives of each student showing "before" and "after" pictures. Less expensive is the silhouetteograph which gives "positives" instead of "negatives."[8] These pictorial records of a student's posture should be kept with the physical inspection record and should be discussed with the student. On the picture should be placed a pencil dot just over the ankle joint, just in front of the knee joint, just back of the hip joint, through the tip of the shoulder and just back of the ear; these dots should be connected by a continuous line which will be a straight line if the posture is correct, i.e., if the lines of pull of gravity are correct.

Warning

Never allow a student to leave a class or conference with a posture picture in her possession. These records should be collected at the end of each conference and returned to the file and kept in a locked place. Serious difficulties may easily arise from carelessness in this matter.

Skeleton

If time permits and a skeleton is available it should be used. Observation of the bony framework of the body will give a better understanding of the need for correct alignment of body segments.

Student Record Sheets

Each student should be given a record sheet to keep in her locker so that it will be at hand when needed for classwork. The days when the sheets will be needed should be announced ahead of time so that time will not be wasted in sending pupils back to the dressing room to procure them.

Figures 70 and 71 show the front and rear views of such a student record form. On the body outlines depicted the student should check the one that most nearly matches her own posture picture, placing check marks at the parts where posture defects need the most attention.

MOTIVATING THE PUPILS

Interest in the fundamentals course can be aroused easily through such devices as contests, the testing program itself, tag days, and the organization of a planning board or some

[8]For information on this machine see C. H. Hubbard: The Silhouetteograph. J. Health & Phys. Educ. 5:16 (March) 1934 and Lee: op. cit., p. 185.

Name.................................... Semester........ Year........

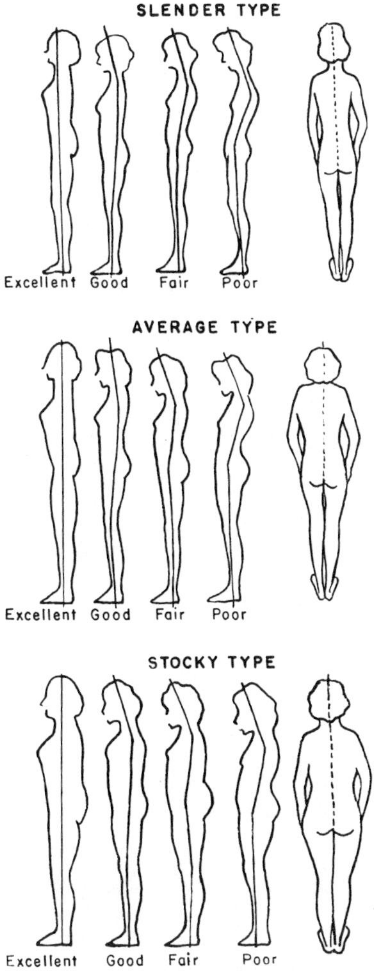

Instructions

1. Underscore the type to the left that describes you the best.
2. In that type group underscore the adjective that best describes your posture as compared with your silhouetteograph.
3. On that body outline check the parts that most need correction.
4. List below the faults which you most need to work on for improvement:
   a. Posture
      (1)
      (2)
      (3)
      (4)
      (5)
   b. Body mechanics
      (1)
      (2)
      (3)
      (4)
      (5)
5. Record your weight

Date	What You Should Weigh	What You Do Weigh
1.		
2.		

Fig. 70. Student work sheet—front view.

other such form of student participation in the planning and conduct of the course.

Contests

Everyone enjoys a contest. Each main topic of the course can readily be worked into one. Each class can set a day when it will select its girl with the best standing posture and

PRESCRIPTION OF EXERCISES

Conditioning Exercises

No.	Name of Exercise	Purpose of Exercise
1.		
2.		
3.		
4.		
5.		
6.		
7.		
8.		
9.		
10.		
11.		
12.		

Special Exercises

1.	For	9.	For
2.	For	10.	For
3.	For	11.	For
4.	For		
5.	For		
6.	For		
7.	For		
8.	For		

Endurance Exercises

1.	
2.	
3.	
4.	

Fig. 71. Student work sheet—rear view.

the ones who walk, sit, climb, run, jump, stoop, and carry the most efficiently. The critical analysis which the students will bring to these final selections will be in itself a fine educational experience if the teacher will guide the experience into considerations of the basic principles involved. The contest itself will spur most on to better performance.

Besides these in-class contests it is easy to inspire out-of-class work on endurance exercises through long hikes and bike trips, cross-country runs, and "road work" over the weekends. Those who enter these contests should sign up with the teacher and then should turn in weekly reports of their contest activity. Suggestions for such contests are given in Chapter VII.

Warnings on Use of Cross-Country Running and "Road Work"

Cross-country running means running through woods, meadows, open fields, parks, and the like, not along streets and roads. "Road work" means alternate brisk walking and running (not jogging) on streets, roads or any other place where the distance can be covered as quickly as possible without encountering obstacles. As a rule, girls' physical education departments do not offer these activities, but they are excellent endurance builders and there is no reason why they should not be used if common sense is employed in planning the distances and if caution is observed in the lead-up as well as actual runs. The U. S. Office of Education has issued warnings for this work[9] which the authors have adapted and enlarged upon as follows:

1. No girl or boy should be permitted to take part in cross-country running or "road work" who has not been pronounced in good physical condition by a properly qualified physician.

2. Pupils who are less mature physically than their classmates should not be permitted to run as long distances as the others.

3. For preliminary "road work," pupils should practice at short bursts of speed and jogging on the grass.

4. Girls should not be permitted to run even moderate distances at top speed until they have gone through at least two weeks of practice at short bursts and jogging.

---

[9]United States Education Office, F.S.A.: Physical Fitness Through Physical Education for Victory Corps. Pamphlet No. 2, 1942, pp. 30-31 and 54.

5. In "road work," the distance to be covered in the early stages should be no more than one to two miles with the walking stretches longer than the runs. For example, in a one-mile run which covers twelve standard city blocks early "road work" would mean brisk walking for three blocks and running for one block, repeating this ratio of walking and running until the distance is covered. Gradually the walking distance is decreased and the running increased. Undue breathlessness will warn the teacher when she has stepped up the running distance or the speed too rapidly.

6. Progression in "road work" is attained not only by decreasing the walking distances and increasing the running distances but also by decreasing the time of the performance. This progression should not be so rapid, however, that it puts a strain on the heart-lung mechanism as evidenced by undue breathlessness or fatigue from which the pupil does not recover within an hour.

7. For girls, start with one mile for "road work" and work up to no more than four or five miles at the most, and this only for the more sturdy. Three miles is a good maximum for the average girl.

8. On cool days have the girls wear slacks and light-weight sweaters instead of gym suits.

Student Participation in Planning the Course

Any teacher who hopes to procure from her pupils an intelligent interest in the class plans rather than blind, docile acceptance will do well to formulate a committee or board of carefully selected, representative students with whom she can discuss the plans of the course, giving them an opportunity to ask questions, to suggest changes, and giving her, the specialist, an opportunity to explain plans to them, the amateurs. The U. S. Office of Education defines wisely the functions of such student planning groups when it advises the "use of selected students to help build curriculum, **not as experts,** but as interpreters of student opinion and future participants."[10] It is,

[10]From leaflet, *Adjustment of the College Curriculum to Wartime Conditions and Needs*, Part II, "Function of Physical Education in American Democracy," p. 2. The bold face type used in the quoted material is not in the original. They are used here for added emphasis.

indeed, important that students understand that they are not competent to plan the course alone but that they are welcome to make suggestions and to ask questions about it for their own better understanding and for their more effective interpretation to the other students. Also they should feel free to inform the teacher of student reactions which may suggest that she is pushing the class too rapidly or too slowly in certain parts of the course, and they should not hesitate to appraise her of unfavorable student attitudes toward the course which may escape her notice and may need her attention.

A Venture In Democracy

During the school year 1942-43, the Department of Physical Education for Women at the University of Nebraska organized a Planning Board made up of one representative chosen by the students from each of the thirteen sections of the freshmen fundamentals course. The published report of this venture was in part as follows:

> The schools should send children out into the world with an education in physical training which can sustain them all the rest of their lives, not only in their need for recreation, but also in their need for the knowledge and practice of exercise as it applies to fitness. Why they do not do this is probably due to reasons such as the following: (1) too few teachers are properly qualified to pass on this knowledge; (2) the schools do not permit sufficient time for such training; (3) too many teachers who are qualified do not give the extra thought and energy needed to add to their classwork that extra something that gives it carry-over value; (4) too few teachers let the pupils in on the secret of what the program is all about; and (5) too few teachers let the pupils help plan the work so that they may have a better understanding and appreciation of it. The first of these faults can be remedied by more and better professional training to reach an ever-increasing number of trainees; the second, by training of school men for a better understanding of the needs of children in their physical education; the third, by eliminating "clock punchers" from the profession as quickly as possible; the fourth and fifth, by a venture in democracy. It is this venture which I want to discuss.
>
> We talk enthusiastically about our democratic form of government, and we are willing to defend it even with our lives, but how deeply do we believe in the democratic procedure for our educational institutions, particularly for our own part of the school program? The democratic process is usually a slow process so that it requires the patience of a Job for the efficiency-expert type of teacher to carry on his work in a really democratic way, but that way pays high dividends in satisfactions gained from group effort and in lasting impressions that come from slow but steady growth . . .
>
> Early in the year all classes met together in two large mass meetings where the Director talked to the hundreds of girls enrolled in the department about

their need for physical fitness and what we hoped to do to help them achieve it. The purpose back of our classwork and the why and wherefore of the type of work we proposed to give were explained to them.

After classwork had been carried on for some time so that the girls had come to know each other, each class was invited to elect one member as its official representative to sit on a Planning Board which would work with the staff on fundamental plans for future classwork . . .

The students elected to the Planning Board took the responsibility seriously and, to our surprise, looked upon the assignment as an honor. At the start the Board met once a week (this at the request of the students themselves) but it soon developed that this demanded too much time and it was decided to meet twice a month at the beginning of a term and only once a month after that.

The representatives reported back to their classes on Board meetings and called for open discussion of questions that had proved controversial to the board itself. In this way students felt free "to get it off their chests" and many availed themselves of the privilege. They told the staff members some very frank and interesting things, and the staff had an opportunity to answer charges and explain policies. Frequently class members charged their representatives with the responsibility of presenting certain topics to the Board for discussion and asked that a report on the matter be returned to them. Constructive suggestions far outweighed unfavorable criticisms. The experience as a whole was one of friendly cooperation.

Through this Planning Board a very definite feeling of comradeship in a joint effort was built up between students and staff, and the staff felt that there was a decided increase in interest in classwork on the part of the students. Perhaps the most interesting outcomes of this venture were the following student requests for this coming year: (1) more strenuous work (they seem to resent it if faculty inspired but like to prescribe it for themselves); (2) revision in the type of final examination given . . . ; (3) a chance for all to learn to swim; (4) a more gradual lead-up to strenuous work than we gave last fall; (5) a daily "work-out" of some sort beyond the 3-hour-a-week class requirement; (6) a continuance of the Planning Board with representatives from each class; . . . (7) more explanations from class instructors as to the why of the particular form of presentation of classwork.

Although this democratic method slowed down the class progress in some ways it aroused such interest in the work and such intelligent questionings and criticisms on the part of the students that the gains more than offset the losses . . . [11]

Through the years of the existence of this Freshmen Planning Board the students have offered many suggestions and criticisms, a few of which are as follows:

1. More musical accompaniment for rhythmical work.

2. Calisthenics kept up for entire year, not just for the first semester.

3. More strenuous work for some sections where the students felt they weren't being given as strenuous work as in other sections.

---

[11]Mabel Lee: A Venture in Democracy. J. Health & Phys. Educ. 14:366 (September) 1943.

4. Use of swimming once a week over a long period of time as a supplement to other work rather than offering it three times a week intensively.

5. The knowledge tests should carry less weight than was given them. (The instructors are not ready to concede this point.)

The Director of the Department, who has no opportunity to teach any of the freshmen sections herself, considers this Planning Board invaluable for it gives her close contact with the students through their chosen representatives and gives her an excellent opportunity to "keep her ear close to the ground" about the course. All instructors who teach a section of the course are invited to sit in on board meetings, but, whether they are present or not, a full report of all meetings is given to all instructors. Frequently student suggestions are such that the Director calls the staff together to discuss these suggestions, and if the majority thinks it wise, changes are made in the course or its administration to meet the student recommendations. If the changes are not to be made, an explanation of why not is reported back to the Board.

CHAPTER XII

LESSON PLANNING

The amount of material which can be covered in a single class period or in a semester will depend upon the number of times the class meets, the length of each class period, and the experience and skill of the teacher. The types of material to be included will depend upon the needs of the students. Therefore no one individual can make out a set of lesson plans which will fit all situations. The lesson plans outlined in this chapter are designed merely to show how the various items which might be included can be worked into a total course. Professionally trained and experienced teachers probably prefer to make their own lesson plans, but untrained or inexperienced teachers might do well to use or adapt the lesson plans given here.

Information on the preliminary work which needs to be done on course outlines before the course calendar and detailed lesson plans are made has been discussed in detail in Chapter XI.

This chapter will show how a course calendar is made up from material in a course outline, and then how this course calendar is expanded into complete lesson plans.

CONSTRUCTION OF THE CALENDAR

Before starting work on construction of the calendar, it would be well to decide how many lessons should be spent on each main topic in the outline, how many on testing and reviews. These should be totaled and the final sum compared to the actual number of class periods. If there are not enough periods, then decisions must be made as to what shall be pruned down and what, if any, deleted, until the two total figures match.

For a sample course calendar see Table V. When the calendar is completed with each class period filled in and the contents of the course all written in, each into its proper niche, then the work on lesson plans is ready to start.

LESSON PLANS

The usual lesson plan will function more smoothly when it has been clearly and definitely worked out in great detail beforehand. The more inexperienced the teacher the more detailed should be the preliminary planning. Only the expert and experienced teacher may safely indulge in impromptu teaching, and even then she should attempt it only at rare intervals.

The lesson should be so planned that the pupils will be aware of a gradually increasing intensity of muscular work, a pleasant sense of fatigue, and a feeling of well-being. If undue fatigue or strain results the lessons have not been adequately prepared for gradual progression from one lesson to another of intensity of muscular work, or else there has been inadequate attention to warming-up with a gradual rise to a peak of muscular effort and a gradual decrease toward the end of each class period.

Within the class hour while progressing from easy to more difficult exercises, the teacher should not blindly cling to her advance plans should she note that the class apparently is not ready to progress as rapidly as hoped for. Constant adjustments will need to be made as the actual classwork unfolds, but this does not excuse the teacher from making plans.

Guiding Principles

In planning the lessons step by step there are certain guiding principles such as the following:

1. Start any and all exercise periods with a slow, easy run (not speed running) or other warm-up work such as rope jumping or flexibility movements. From this build up gradually to the heavier work requiring strength and then taper off again.

2. Plan for a slow but steadily increasing demand of work in each class lesson. This means a carefully thought out and planned dosage of actual number of repetitions of each exer-

TABLE V

COURSE CALENDAR

Monday	Wednesday	Friday
Sept. 25: Report in street clothes Roll call Announcements of departmental rules and regulations	Sept 27: Body-mechanics test	Sept. 29: Learn short method roll call Content and objectives of course Start learning conditioning series Shower procedure and regulations
Oct. 2: Finish learning conditioning series Partners for correction	Oct. 4: Discussion of need for fitness Correct running Relaxation	Oct. 6: Discuss results body-mechanics test and silhouettographs (if taken)
Oct. 9: Continue discussion of body-mechanics test results	Oct. 11: Review relaxation Discuss and practice correct standing Discussion and practice of endurance exercises	Oct. 13: Review relaxation Discuss overload principle Discussion and drill on pelvic tilt
Oct. 16: Review correct standing and pelvic tilt Discussion of and exercises for sacro-iliac region Review relaxation	Oct. 18: Review correct standing and pelvic tilt Review exercises for sacro-iliac region Discussion and practice of exercises for abdomen and lower back	Oct. 20: Score on Full-Squat test Review exercises for sacro-iliac and abdomen and lower back

TABLE V (Continued)

Monday	Wednesday	Friday
Oct. 23: Score on Sit-Up test Review exercises for abdomen and lower back Discuss scapulae and shoulder position	Oct. 25: Score on Wing-Lift test Review exercises for lower back and abdomen Discussion and exercises for forward shoulders and flat chest	Oct. 27: Score on Push-Up test Review exercises for forward shoulders and flat chest Discuss feet
Oct. 30: Review exercises forward shoulder and flat chest Discussion and exercises for feet	Nov. 1: Review exercises for feet and correct standing Discuss and demonstrate foot position and appearance	Nov. 3: Review exercises for feet Discussion and demonstration of shoes
Nov. 6: Review exercises for feet Hygiene of feet and hosiery Influence of heels on posture Tests of body alignment	Nov. 8: Review exercises for feet, body alignment test, and correct standing Discussion-demonstration lying positions and everyday activities	Nov. 10: Review exercises for feet and for abdomen and lower back Discussion-demonstration of walking
Nov. 13: Review exercises for feet and abdomen and lower back Discussion of and practice drills in walking	Nov. 15: Review correct standing, exercises for forward shoulders and flat chest, and walking drills Discuss walking for long distances	Nov. 17: Review practice drills for correct walking and exercises for abdomen and lower back Discuss menstruation

TABLE V (Continued)

Monday	Wednesday	Friday
Nov. 20: Review correct standing and exercises for abdomen and lower back Midsemester written test	Nov. 22: Push-Up and Full-Squat strength test Discuss menstruation (cont.)	Nov. 24: Sit-Up and Wing Lift strength test Review correct standing and pelvic tilt and relate to pelvic organ support Discussion and practice of exercises for dysmenorrhea
Nov. 27: Discussion of dysmenorrhea (cont.) Review knee-chest position Learn Billig exercises Discuss constipation and relate to above	Nov. 29: Review exercises for dysmenorrhea Discussion and exercises for constipation Discuss and practice jumping Discuss smoking and exercise	Dec. 1: Review exercises for dysmenorrhea and constipation, and jumping New jumping drills Discuss smoking (cont.)
Dec. 4: Review Billig exercises and jumping Review exercises for forward shoulders and flat chest Discuss and practice correct sitting and rising	Dec. 6: Review jumping, sitting, rising, and exercises for sacro-iliac region Discuss proper chairs and tables Discuss and practice stooping Discuss lifting and carrying	Dec. 8: Review correct sitting, rising, stooping Discuss and practice lifting and carrying various articles
Dec. 11: Review Billig exercises and exercises for feet Review practice lifting and carrying	Dec. 13: Review exercises for feet Discussion and practice of climbing Discuss weight control	Dec. 15: Discuss (cont.) and review climbing Discuss weight control (cont.)

TABLE V (Continued)

Monday	Wednesday	Friday
Dec. 18: Review climbing Discuss and practice reaching and pushing	Jan. 3: Discuss effective use of body in home-making and industry Organize class for review	Jan. 5: Review Groups I, II, III
Jan. 8: Review groups IV, V, VI	Jan. 10: Review groups VII, VIII, IX	Jan. 12: Final check on strength tests Discuss results
Jan. 15: Body-mechanics test	Jan. 17: Discuss results of all tests	Jan. 19: Written examination

cise in the conditioning series each day, and a gradually increasing demand of work of the whole class period. The less vigorous elements of body mechanics should be taught and practiced at the beginning of the semester and the more vigorous at the end. Some muscular discomfort or stiffness is sure to result when starting an activity program after a long period of inactivity. Extreme discomfort, however, is not necessary nor desirable, and a carefully planned dosage of work will prevent it. Students should be taught to understand this so they will know what to expect. Those who suffer extreme discomfort should be singled out for personal conference.

3. For the "warmer-uppers" and "conditioners" of each class hour use the same few exercises over and over—these few to be carefully chosen for what they are supposed to do, with progression attained day by day by increasing the number of repetitions or by increasing the intensity. No teacher need apologize for the meagerness of her quantity offering in "warmer-uppers" and "conditioners": it is quality here that counts.[1] The teacher who boasts of the great variety of exercises taught her classes for conditioning falls short of her own good opinion of herself, for she has probably succeeded merely in giving her pupils educational indigestion.

4. Plan sufficient time for endurance exercises in each lesson to make a real demand upon the heart and lungs.

5. Place in each lesson some new educational material so that the course may progress on towards its conclusion.

6. In each lesson devote some time to a review of the previous topics presented. Be sure that each topic is presented at least three times, once for introduction and at least twice for review.

7. Integrate maxims of healthful living with every lesson of the exercise program.

8. Keep attuned to the universal needs and urges of pupils and coordinate class activities with these urges. For example, every girl wishes to be attractive, therefore coordinate this desire with the importance of good posture in attaining the desire.

---

[1]This point of view has been more fully discussed in Chapter IV. It is offered again merely as a reminder of a topic well worth a reminder.

9. Build lessons around the idea that although physical fitness is not an end in itself as a life goal it is an important factor toward attaining any goal one sets for herself.

10. Utilize every opportunity for education. Every exercise and every activity furnishes an opportunity for practice in effective movement and aids in establishing correct habits.

11. Teach pupils to analyze all postures or movements in terms of the basic principles governing correct body mechanics. Instead of telling them how to do a task effectively, let them figure it out.

12. Plan all parts of each lesson so that all pupils will be kept busy all the time.

13. Allow time for testing, reviewing, and final examinations.

A Complete Lesson Plan

A well-organized and complete lesson plan should include more details than the sample lessons which follow in this chapter. For example, the first part of the outline should include a category called "equipment to have in readiness." Under this should be listed everything necessary for the class hour, notices to be posted, equipment to be assembled, charts to have ready, or score sheets to prepare.

Where the following lessons merely state "Supplementary exercises for ——" a complete lesson plan should indicate the name or describe briefly what these exercises are.

It is also wise to outline briefly the main points to be covered in discussion so there will be less tendency to digress and so one can be sure of having covered all important points.

Skeleton Lesson Plans

Skeleton lesson plans for a class which meets three times a week for fifteen weeks are outlined below. Each class period is approximately forty-five minutes in length.

LESSON 1

1. Class reports in street clothes.
2. Roll Call.
3. Discuss department rules and regulations.
4. Announce gymnasium costume requirements.
5. Announce all to report next class hour in either a bathing suit or shorts and brassiere ready for the body-mechanics test.
6. Other details, such as locker assignments.

LESSON 2

1. Class in bathing suits or shorts and brassiere.
2. Body-mechanics test (outline the details of procedure).

LESSON 3

	Minutes
1. Class in regular gymnasium costume.	
2. Learn short method of taking roll	3½
3. Warmer-Upper run	1½
4. Brief discussion of content and objectives of course and the place of this course in physical education and education	8
5. Learn how to open order for exercises	2
6. Name, demonstrate, tell main purpose of conditioning exercise, Number 1	2
7. Class execution of the exercise minimum number of times	2
8. Repeat 6 and 7 for the conditioning exercises, Numbers 2 through 7.	23
9. Brief discussion of muscle soreness, its meaning, its symptoms, and its cause	2
10. Announcement of shower procedures and regulations	1
Total	45

LESSON 4

	Minutes
1. Roll Call	1½
2. Warmer-Upper run	1½
3. Open order for exercises	1
4. Arrange class in partners for correction of execution of conditioning exercise series	1
5. Review conditioning exercise, Number 1; state its purpose briefly, demonstrating while talking	1
6. One of each couple practice the exercise with partner help and correction	1
7. Second of each couple repeats	1
8. Repeat 5, 6, and 7 for each conditioning exercise, Numbers 2 through 7, learned last class period	21
9. Name, demonstrate, tell main purpose of next new conditioning exercise, Number 8	2
10. One of each couple practices the exercise with partner help and correction	1
11. Second of each couple repeats	1
12. Repeat 9, 10, and 11 for each of the next two exercises, Numbers 9 and 10	8
13. Name, demonstrate, tell main purpose of next two conditioning exercises, Numbers 11 and 12, and class practices all at once (partners not necessary)	4
Total	45

LESSON 5

	Minutes
1. Roll call	½
2. Warmer-Upper run	1½
3. Open order	1
4. Conditioning exercises, each executed number of times decided upon for second week (partner correction for Push-Ups only)	21
5. Discussion topics	10
a. Need for fitness.	
b. Why conditioning exercises.	
c. Running.	
(1) Correct technique.	
(2) Demonstration.	
6. Class practice correct running	1
7. Second discussion topic: Relaxation	5
a. Need for learning how.	
b. Can be acquired by practice.	
8. Class practice of relaxation techniques	5
Total	45

LESSON 6

	Minutes
1. Roll call	½
2. Warmer-Upper run (emphasis on correct form)	1½
3. Open order	1
4. Conditioning exercises (partner correction for Push-Ups only)	20
5. Discuss body-mechanics test (and silhouetteograph if taken) in general	10
a. Purpose of taking both.	
b. Difference between the above and medical examination given by a physician.	
6. Pass out results of body-mechanics test and silhouetteograph (if taken) to each pupil	11
a. Explain how to interpret findings by introducing pupils to basic principles for correct body mechanics.	
b. Explain difference between faulty posture and faulty body mechanics in terms of basic principles.	
7. Collect all material and announce that this will be shown again	1
Total	45

LESSON 7

	Minutes
1. Roll call	½
2. Warmer-Upper run (emphasis on correct form)	1½
3. Open order	1
4. Conditioning exercises (partner correction for Push-Ups only)	20
5. Pass out material showing results of body-mechanics test and silhouetteographs (if taken)	21
a. Answer questions.	
b. Class discussion of principles related to the test.	
c. Explain again how faults will be corrected.	
6. Collect cards and silhouetteographs	1
Total	45

LESSON 8

	Minutes
1. Roll call	½
2. Warmer-Upper run (emphasis on correct form)	1½
3. Open order	1
4. Conditioning exercises (partner correction for Push-Ups only)	20
5. Review topic: Relaxation.	
a. Preliminary talk	1½
b. Review practice techniques of preceding class period	2
c. Practice new techniques	3
6. First topic of day: Correct Standing.	
a. Preliminary talk on attitudes and appreciations	5
b. Discussion of principles related to standing	3
c. Demonstration and practice of standing tall with wide shoulders	1½
7. Second topic: Endurance.	
a. Preliminary talk on heart-lung endurance and muscular endurance	3
b. Class does Sit-Up exercise until symptoms of muscular fatigue are experienced	1½
c. Class runs around room. (All are urged to run daily out of class until "second wind" is experienced.)	1½
Total	45

LESSON 9

	Minutes
1. Roll call	½
2. Warmer-Upper run (emphasize correct form)	1½
3. Open order	1
4. Conditioning exercises (partners for Push-Ups)	19
5. Review topic: Relaxation.	
a. Review practice previous techniques	5
b. Practice for total complete relaxation	4
6. Second review topic. Endurance.	
a. Talk on overload principle	4
b. Class practice of overload principle by increasing intensity and speed of exercises	4
7. New topic of day: Pelvic Tilt.	
a. Preliminary discussion of basic principle involved	2
b. Demonstration and drill on practice techniques	4
Total	45

LESSON 10

	Minutes
1. Roll call	½
2. Warmer-Upper run (correct form)	1½
3. Open order	1
4. Conditioning exercises	17
5. Review correct standing and pelvic tilt.	
a. Additional information	2
b. Standing tall	1
c. Pelvic-Tilt practice	2
6. Topic of day: Sacro-iliac Strains.	
a. Preliminary talk on causes	5
b. Demonstration and practice of exercises for ligaments that are too unyielding	5
7. Review and practice total relaxation	10
Total	45

LESSON 11

	Minutes
1. Roll call	½
2. Warmer-Upper run	1½
3. Open order	1
4. Conditioning exercises	17
5. Review practice.	
Standing tall	1
b. Pelvic-Tilt practice	2
c. Practice of exercises for unyielding ligaments in sacro-iliac region	3
6. Topic of day: Lower Back and Abdominal Region.	
a. Preliminary talk on spine, lower back, abdomen, and pelvic-tilt structure and function	7
b. Demonstration, explanation, and practice of supplementary exercises for lower back and abdomen	12
Total	45

LESSON 12

	Minutes
1. Warm-Up run (not for endurance)	1
2. Take roll while pupils pick up individual strength-test score cards and pencils	1
3. Pupils score each other on the Full-Squat test	15
4. Pupils return to roll call numbers and either place cards on correct numbers or teacher collects them in order	1
5. Open order for exercises	1
6. Conditioning exercises (omitting Full Squats)	15
7. Review practice.	
a. Sacro-iliac exercises	2
b. Supplementary exercises for lower back and abdomen	6
8. Supplementary information on lower back and abdomen including class discussion of basic principles underlying good position	3
Total	45

LESSON 13

	Minutes
1. Run for warm-up only	1
2. Take roll while pupils pick up individual strength-test score cards and pencils	1
3. Inform class that any who are menstruating and are in pain or who find exercise causes discomfort are not to do the Sit-Up test today but make it up at a later date	1
4. Pupils score each other on Sit-Up test	15
5. Return to roll-call numbers and collect cards and pencils	1
6. Open order for exercises	1
7. Conditioning exercises (omitting Sit-Ups)	15
8. Review practice.	
a. Supplementary exercises for lower back and abdomen	5
9. Topic of day: Scapulae and Shoulder Position.	
a. Preliminary talk	5
Total	45

LESSON 14

	Minutes
1. Run for warm-up only	1
2. Take roll while pupils pick up individual strength-test score cards and pencils	1
3. Pupils score each other on Wing-Lift test	10
4. Return to roll-call numbers and collect cards and pencils	1
5. Open order for exercises	1
6. Conditioning exercises (omitting Wing Lifts)	15
7. Review practice.	
a. Supplementary exercises for lower back and abdomen	5
8. Topic of day: Head, Shoulders, and Chest.	
a. Preliminary talk with class discussion of basic principles applying here	5
b. Demonstration, explanation, and practice of two of supplementary exercises for forward shoulders and flat chest	6
Total	45

LESSON 15

	Minutes
1. Run for warm-up only	1
2. Take roll while pupils pick up score cards and pencils	1
3. Pupils score each other on Push-Up test	10
4. Discussion of test ratings and meaning of test results	3
5. Return to roll-call numbers and collect cards and pencils	1
6. Open order	1
7. Conditioning exercises (omitting Push-Ups)	16
8. Review practice.	
a. Supplementary exercises for forward shoulders and flat chest learned last class	4
9. Topic of day: Feet.	
a. Preliminary talk on structure and function	8
Total	45

LESSON 16

	Minutes
1. Roll call	½
2. Warmer-Upper	1¾[2]
3. Open order	1
4. Conditioning exercises	16¾
5. Review practice.	
a. Supplementary exercises for forward shoulders and flat chest	4
6. Topic of day: Feet.	
a. Talk on structure and function (continued)	8
b. Demonstration and drill on assuming correct foot position	2
c. Demonstration and practice of supplementary exercises for the feet	8
d. Discussion of value of exercise for feet	3
Total	45

[2]It may seem to some that it is impossible to check as small an amount as fifteen seconds but the dosage of running (or rope jumping) for warming up and endurance should be increased by fifteen-second intervals. Since several years' experience has proven that the first four items in the lesson plan can almost always be covered in twenty minutes, this is not an unreasonable request. One item increases only at the expense of another, hence the quarter minute increase or decrease in time intervals.

LESSON 17

	Minutes
1. Roll call	½
2. Warmer-Upper	1¾
3. Open order	1
4. Conditioning exercises	16¾
5. Review practice.	
a. Supplementary exercises for feet	6
b. Correct standing practice with partner correction and discussion of principle related to habit formation	2
6. Topic of day: Foot Position and Appearance.	
a. Preliminary talk	5
b. Class participation in demonstration and observation of appearance caused by various foot positions	12
Total	45

LESSON 18

	Minutes
1. Roll call	½
2. Warmer-Upper	2
3. Open order	1
4. Conditioning exercises	16½
5. Review practice.	
a. Supplementary exercises for feet	5
6. Topic of day: Shoes.	
a. Preliminary talk to develop attitudes and appreciations	5
b. Display demonstration and discussion of good and poor features of currently popular shoes	10
c. Discussion of street shoes of class members and how they reflect foot posture	5
Total	45

LESSON 19

	Minutes
1. Roll call	½
2. Warmer-Upper	2
3. Open order	1
4. Conditioning exercises	16½
5. Review.	
a. Supplementary exercises for feet	5
6. First topic of day: Hygiene of the Feet, Posture, and Heels.	
a. Preliminary talk on hygiene of the feet	5
b. Demonstration of influence of various height heels on posture with discussion of principles involved	5
7. Second topic: Tests of Posture.	
a. Demonstration and practice of wall•test of posture	5
b. Demonstration and practice of vertical line test of body alignment	5
Total	45

LESSON 20

	Minutes
1. Roll call	½
2. Warmer-Upper	2
3. Open order	1
4. Conditioning exercises	16½
5. Review.	
a. Supplementary exercises for feet	5
b. Wall practice of body alignment	2
c. Pupils practice acquiring "muscle feel" of good posture	3
6. Topic of day: Everyday Activities and Posture.	
a. Discussion of basic principles for correct body mechanics in everyday activities, including demonstration, and questions	15
Total	45

LESSON 21

	Minutes
1. Roll call	½
2. Warmer-Upper	2
3. Open order	1
4. Conditioning exercises	16½
5. Review practice.	
a. Supplementary exercises for feet	5
b. Supplementary exercises for abdominal and lower back region	5
6. Topic of day: Walking.	
a. Preliminary talk	5
b. Demonstrations of correct and incorrect walking with discussion by class of principles involved	10
Total	45

LESSON 22

	Minutes
1. Roll call	½
2. Warmer-Upper	2
3. Open order (emphasis on correct walking)	1
4. Conditioning exercises	16½
5. Review practice.	
a. Supplementary exercises for feet	5
b. Supplementary exercises for abdomen and lower back	5
6. Topic of day: Walking (continued).	
a. Preliminary talk to develop appreciation	5
b. Practice drills in walking, including slow-motion walk	10
Total	45

LESSON 23

	Minutes
1. Roll call	½
2. Warmer-Upper	2
3. Open order (emphasis on correct walking)	1
4. Conditioning exercises	16½
5. Review practice.	
a. Correct standing position (with correction)	2
b. Supplementary exercises for forward shoulders and flat chest	4
c. Practice drills for correct walking	9
6. Topic of day: Walking for Long Distances.	
a. Information on techniques for long hikes	10
Total	45

LESSON 24

	Minutes
1. Roll call	½
2. Warmer-Upper	2¼
3. Open order (emphasis on correct walking)	1
4. Conditioning exercises	16¼
5. Review practice.	
a. Practice drills for correct walking	3
6. Topic of day: Menstruation.	
a. Preliminary talk on menstruation as a normal function	7
b. Structure of pelvic organs (use anatomy charts and, if available, a manikin)	10
7. Review supplementary exercises for abdomen and lower back and emphasize importance in relation to menstrual function	5
Total	45

LESSON 25

	Minutes
1. Roll call	½
2. Warmer-Upper	2¼
3. Open order	1
4. Conditioning exercises	16¼
5. Review practice.	
a. Correct standing, including pelvic tilt	2
b. Supplementary exercises for abdomen and lower back	5
6. Written test covering knowledge, understanding, and appreciations of material of conditioning exercises, supplementary exercises, basic principles for correct body mechanics, and discussion topics	18
Total	45

LESSON 26

	Minutes
1. Warm-up run (not for endurance)	1
2. Take roll while pupils pick up individual strength-test score cards and pencils	1
3. Pupils are arranged in partners and score each other on the Push-Up and Full-Squat tests	15
4. Pupils return to roll-call numbers; cards and pencils are collected	1
5. Open order for exercises	1
6. Conditioning exercises (omitting Full Squats and Push-Ups)	13
7. Topic of day: Menstruation (continued).	
a. Physiology of menstruation	7
b. Personal hygiene do-and-don't facts	3
c. Relation of exercise to function	3
Total	45

LESSON 27

	Minutes
1. Run for warm-up only	1
2. Take roll while pupils pick up individual strength-test score cords and pencils	1
3. Pupils are arranged in partners and score each other on the Sit-Up and Wing-Lift tests	13
4. Discussion of meaning of test results	2
5. Pupils return to roll-call numbers; cards and pencils are collected	1
6. Open order for exercises	1
7. Conditioning exercises (omitting Sit-Ups and Wing Lifts)	13
8. Review practice.	
a. Correct standing, including pelvic tilt, and hold position while instructor explains relationship of tilt to support of pelvic organs and reminds class again of basic principle related to habit formation	2
9. Topic of day: Dysmenorrhea.	
a. Talk on prevalence and amount of incapacitating illness	3
b. Causes	2
10. Demonstration and practice of exercises for dysmenorrhea:	
a. Knee-chest position	3
b. Alternate leg extension from knee-chest position	1
11. Information on use of abdominal pumping to relieve congestion (an exercise in conditioning series)	2
Total	45

LESSON 28

	Minutes
1. Roll call	½
2. Warmer-Upper	2¼
3. Open order	1
4. Conditioning exercises	16¼
5. Review practice.	
a. Knee-chest position (with and without leg extension)	3
6. First topic of day: Dysmenorrhea (continued).	
a. Treatment or prevention	4
b. Demonstration and practice of Billig exercise	5
c. Answer questions about hygiene and exercise in relation to menstruation and dysmenorrhea	5
7. Second topic of day: Constipation.	
a. Preliminary talk on causes and prevalence	6
b. Relation of constipation to dysmenorrhea	2
Total	45

LESSON 29

	Minutes
1. Roll call	½
2. Warmer-Upper	2½
3. Open order	1
4. Conditioning exercises	16
5. Review practice.	
a. Exercises for dysmenorrhea	3
(1) Abdominal Pumping.	
(2) Knee-chest position.	
b. Billig exercise with information on its exact use to relieve dysmenorrhea, including relapse	3
6. First topic of day: Special Exercise for Constipation.	
a. Types of exercises which help	1
b. Discussion of the four exercises of conditioning series which help	2
c. Demonstration and practice of the supplementary exercises for constipation	5
7. Second topic of day: Jumping.	
a. Preliminary talk on attitude and necessity of learning correct technique	1
b. Discussion of principles relating to correct jumping	1
c. Demonstration and practice of practice drills 1, 2, and 3 under jumping	4
8. Discussion topic: Smoking in Relation to Exercise.	
a. Preliminary talk giving a few facts about smoking	5
Total	45

LESSON 30

	Minutes
1. Roll call	½
2. Warmer-Upper	2½
3. Open order	1
4. Conditioning exercises (review of purpose)	16
5. Review practice.	
a. Exercises for dysmenorrhea	4
b. Exercises for constipation	3
c. Practice drills of previous lesson on jumping	3
6. Demonstration followed by practice of remaining practice drills on jumping	5
7. Discussion topic: Smoking (continued).	
a. Facts and open discussion leading to pupil's decision of whether or not to smoke	10
Total	45

LESSON 31

	Minutes
1. Roll call	½
2. Warmer-Upper	2¾
3. Open order	1
4. Conditioning exercises	15¾
5. Review practice.	
a. Some of practice drills on jumping	2
b. Billig exercise	2
c. Remainder of practice drills on jumping	3
d. Supplementary exercises for forward shoulders and flat chest	4
6. Topic of day: Sitting.	
a. Preliminary talk to develop attitudes and appreciations, include demonstrations of appearance of various sitting postures with discussion of principles involved	2
b. Demonstration and practice (in couples) of correct sitting posture	2
c. Demonstration and practice (in couples) of correct rising from a chair	3
d. Class works on practice drills suggested in Chapter VII on the mechanics of rising from a chair	7
Total	45

LESSON 32

	Minutes
1. Roll call	½
2. Warmer-Upper	2¾
3. Open order	1
4. Conditioning exercises	15¾
5. Review practice.	
a. Exercises for unyielding ligaments in sacro-iliac region	2
b. Jumping drills	3
c. Correct sitting position and rising from sitting position	3
6. First topic of day: Sitting (continued).	
a. Lecture-demonstration of proper chairs and worktables. Refer to basic principles for correct body mechanics. Have both good and poor examples to show. Have pupils enter the discussion while sitting on floor correctly	5
7. Second topic of day: Stooping.	
a. Class analysis of correct technique in stooping in terms of basic principles	1
b. Class practice in stooping	1
c. Practice combining correct sitting, rising, taking a few steps, stooping, returning to chairs, and being seated	2
8. Discussion topic: Lifting and Carrying.	
a. Class analysis of efficient body mechanics in lifting and carrying in terms of basic principles and analysis of dangers of faulty lifting and carrying techniques	8
Total	45

LESSON 33

	Minutes
1. Roll call	½
2. Warmer-Upper	2¾
3. Open order	1
4. Conditioning exercises	15¾
5. Review practice	3
a. Correct sitting.	
b. Correct rising.	
c. Correct stooping.	
6. Topic of day: Lifting and Carrying.	
a. Lecture-demonstration and class discussion of correct technique of lifting and carrying	10
(1) Several books.	
(2) Heavy suitcase.	
(3) Chair, stool, or bench.	
(4) Tray of dishes.	
(5) Bucket of water.	
(6) Various weights of sandbags.	
b. Divide class into six groups and have each group demonstrate one of the above with criticism from the rest of the class in terms of efficiency as set forth in basic principles	12
Total	45

LESSON 34

	Minutes
1. Roll call	½
2. Warmer-Upper	2¾
3. Open order	1
4. Conditioning exercises	15¾
5. Review practice.	
a. Billig exercise	2
b. Supplementary exercises for the feet	5
6. Review topic: Lifting and Carrying.	
a. Discussion-review of important points of correct technique	3
(1) Stooping.	
(2) Stooping and lifting light object.	
(3) Lifting and carrying heavy object with one hand.	
(4) Lifting and carrying heavy object with two hands.	
b. Divide class into six groups and have each practice lifting and carrying objects listed in preceding lesson	15
Total	45

LESSON 35

	Minutes
1. Roll call	½
2. Warmer-Upper	2¾
3. Open order	1
4. Conditioning exercises	15¾
5. Review practice.	
a. Supplementary exercises for the feet	5
6. Topic of day: Climbing.	
a. Preliminary talk about unsafe climbing practices	2
b. Class discussion of correct stair climbing	2
c. Class practice (in couples) climbing stairs	10
7. Discussion topic: Weight Control in Relation to Exercise	6
Total	45

LESSON 36

	Minutes
1. Roll call	½
2. Warmer-Upper	2¾
3. Open order	1
4. Conditioning exercises	15¾
5. Topic of day: Climbing (continued).	
a. Talk on climbing as endurance builder and hip reducer	2
b. Demonstrate and discuss basic principles related to correct ladder climbing	2
c. Class practice, with criticism, climbing ladders of all kinds. Divide into as many groups as have different kinds of ladders. Rotate groups	11
6. Discussion topic: Weight Control (continued) with class participation	10
Total	45

LESSON 37

	Minutes
1. Roll call	½
2. Warmer-Upper	3
3. Open order	1
4. Conditioning exercises	15½
5. Review topic: Climbing.	
a. Brief review of important techniques making for safety	2
b. Divide class into groups for practice in climbing stairs and various kinds of ladders. Groups rotate	15
6. Topic of Day: Reaching and Pushing.	
a. Preliminary talk to develop attitudes and appreciations	3
b. Class discussion and demonstration of correct reaching and correct pushing	3
c. Group demonstrates pushing by moving piano, rest of class criticises in terms of efficiency of effort	2
Total	45

LESSON 38

	Minutes
1. Roll call	½
2. Warmer-Upper	3
3. Open order	1
4. Conditioning exercises	15½
5. Topic of day: Effective Use of the Body	25
a. General tie-up of all preceding information and all basic principles into effective use of the body in homemaking and industry.	
b. Discussion and questions.	
Total	45

LESSON 39

	Minutes
1. Roll call	½
2. Warmer-Upper	3
3. Open order	1
4. Conditioning exercises	15½
5. Presentation of review of following topics in three groups:	25
Group I. Relaxation.	
Group II. Correct Standing, Pelvic Tilt, and Tests of Body Alignment.	
Group III. Lower Back, Abdominal, and Sacro-iliac Regions.	
Total	45

LESSON 40

	Minutes
1. Roll call	½
2. Warmer-Upper	3
3. Open order	1
4. Conditioning exercises	15½
5. Presentation of review of following topics by three groups:	25
Group IV. Forward Shoulders and Flat Chest, Reaching, and Pushing.	
Group V. Feet and Shoes.	
Group VI. Menstruation and Dysmenorrhea.	
Total	45

LESSON 41

	Minutes
1. Roll call	½
2. Warmer-Upper	3
3. Open order	1
4. Conditioning exercises	15½
5. Presentation of review of following topics by three groups: Group VII. Constipation. Group VIII. Sitting, Rising, Stooping, Lifting, and Carrying. Group IX. Jumping and Climbing.	24
6. Announce final testing on strength and endurance next class period	1
Total	45

LESSON 42

Final Testing of Muscular Strength and Endurance

	Minutes
1. Run for warm-up only	1
2. Take roll while pupils pick up individual strength-test score cards and pencils	1
3. Get scores on the following tests. The procedure is outlined in detail in Chapter X a. Push-Up test. b. Full-Squat test. c. Sit-Up test. d. Wing-Lift test.	40
4. If there is any time left discuss the improvement shown in the test scores and answer questions about test results	2
5. Announce body-mechanics test next class period. All pupils to report in bathing suit or shorts and brassiere	1
Total	45

LESSON 43

Body-Mechanics Test

(The exact procedure for all details of administering the body-mechanics test can be found in Chapter X.)

LESSON 44

	Minutes
1. Roll call	½
2. Warmer-Upper	3
3. Open order	1
4. Conditioning exercises	15½
5. Pass out strength-test score cards, results of first and second body-mechanics test, and silhouetteographs taken at beginning and end of the semester (if taken). Demonstrate and discuss meaning of scores in terms of improvement. Questions and discussion by class members	24
6. Announce to report in street clothes for final written test next class period and tell exactly where to report	1
Total	45

LESSON 45

Written Examination

Class reports at designated place in street clothes for written examination. (For details of procedure see Chapter X.)

THE TEACHER'S RESPONSIBILITY

The way a lesson is conducted reflects a teacher's ideals, knowledge, and standards. Each lesson should embody definite educational objectives and the success or failure of the lesson should be measured by how effectively these objectives have been attained. Each lesson should give pupils the utmost in value for their time and effort.

A lesson carefully planned, well balanced, and clearly objective not only interests pupils but receives their enthusiastic cooperation and response. Transcending this, the course as a whole if carefully planned, well balanced, and clear as to its objectives will not only make a valuable contribution to the general education of the pupils but will also materially help in preparing them for effective living. That one who has an efficient body, under control, and who understands health practices related to effective living is the better able to take her place as a citizen in a democratic society.

INDEX